INGREDIENTS OF CHANGE

INGREDIENTS OF CHANGE

THE HISTORY AND CULTURE OF FOOD IN MODERN BULGARIA

MARY C. NEUBURGER

CORNELL UNIVERSITY PRESS
Ithaca and London

First published 2022 by Cornell University Press

Library of Congress Cataloging-in-Publication Data

Names: Neuburger, Mary, 1966– author.
Title: Ingredients of change : the history and culture of food in modern Bulgaria / Mary C. Neuburger.
Description: Ithaca [New York] : Cornell University Press, 2022. | Includes bibliographical references and index.
Identifiers: LCCN 2021020500 (print) | LCCN 2021020501 (ebook) | ISBN 9781501762499 (hardcover) | ISBN 9781501762581 (paperback) | ISBN 9781501762505 (ebook) | ISBN 9781501762512 (pdf)
Subjects: LCSH: Food habits—Bulgaria—History. | Food—Bulgaria—History.
Classification: LCC GT2853.B9 N48 2022 (print) | LCC GT2853.B9 (ebook) | DDC 394.1/209499—dc23
LC record available at https://lccn.loc.gov/2021020500
LC ebook record available at https://lccn.loc.gov/2021020501

❧ Contents

❦ Acknowledgments

While I alone am responsible for the content of this book, a number of people and institutions have helped and inspired me along the way and are deserving of thanks. I would like to thank various units at the University of Texas (UT) that provided me support throughout the research and writing process, namely the Center for Russian, East European, and Eurasian Studies (CREEES), the Department of History, and the College of Liberal Arts. First and foremost, I want to thank the CREEES faculty and staff for providing me with an intellectual community and a place to call my professional home. I especially want to thank Agnes Sekowksi, assistant director of CREEES, for acting as a sounding board and confidante. You keep me sane. I also want to thank the history department and its past three (heroic) chairs—Alan Tully, Jackie Jones, and Daina Berry. I am in awe of your fortitude and grateful for your support and flexibility.

Thinking back to how this project unfolded, it was the CREEES "Food for Thought: Culture and Cuisine in Russia and Eastern Europe" conference in 2014 that first sparked my interest in research and writing on food. It was there that I met and was inspired by so many colleagues across disciplines who have written on food—Ronald LeBlanc, Stephen Bittner, Adrianne Jacobs, José Alaniz, Anastasia Lakhitova, Katrina Kollegaeva, Nikolai Burlakoff, François-Xavier Nérard, Dragan Kujundzic, Laura Goering, Ana Tominc, and our own student Abbie Weil (who always makes me laugh). Perhaps most importantly I met or reconnected with several Bulgarian colleagues (or scholars who work on Bulgaria), Stefan Detchev, Marcus Wien, Magdalena Slavkova, and Yusong Jung, whose work and collegiality have been important to this project. A special thanks to Stefan Detchev, who has shared his own large body of published work on food history, read all my draft chapters and offered comments, and shared many a delicious meal and bottle of wine in Sofia (and Austin). At a later food conference in Warsaw, I was lucky enough to also meet Albena Shkodrova and Rayna Gavrilova (whom I met with again in Sofia). I drew heavily on their pioneering research.

It has been exciting to see Bulgarian food studies blossom, with the work of these and other scholars. I am hopeful that my work will offer a solid contribution to this new field.

I also want to thank the UT Slavic Department's Keith Livers, who worked diligently with me and some of the scholars from the UT food conference to put together a special issue of *Gastronomica: The Journal of Critical Food Studies*, titled "Culinary Revolutions: Food, History, and Identity in Russia and East-Central Europe," which came out in 2017. Thank you Keith for working with me on the introduction to this issue, for our many conversations on food, and your enlightening article on food in contemporary Russian literature. We benefited from reading and editing the work of our contributors from the 2014 conference—LeBlanc, Weil, Goering, Nérard, Kollegaeva—but also Andrew Klobier, whose work on East German coffee sparked my interest at the annual convention of the Association for Slavic, East European, and Eurasian Studies (ASEEES) in 2016. I enjoyed working with all of you, and *Gastronomica* editor Melissa Caldwell, on this project. Melissa was extremely helpful in pulling this issue together, and her own pre-introduction to our introduction was brilliant! Her work on Russian food is an inspiration.

Another source of support for this project was the UT Department of History's amazing Institute for Historical Studies. My semester of leave through the institute, in the year when the theme was "History of Food and Drink," came at the best possible time for my project. Special thanks to the then director, Seth Garfield, and all the fellows (internal and external) for their feedback on my project that year. Special thanks to Michelle King and (frequent institute visitor) Rachel Laudan, who offered feedback on an early chapter of the book. The work of both of you on food history has been especially valuable for my thinking on food.

I also want to thank so many others at UT for their encouragement and many conversations on food that inspired me to keep writing. Thank you Christian Hilchey for sharing your passion for food, and your amazing cooking! I always enjoyed visiting your course "Cuisine and Culture in Eastern Europe" and discussing my work with your students. I am also grateful for the friendship and scholarly collaboration of Kiril Avramov and Jason Roberts, who have moved forward my thinking on so many issues. Kiril's collaboration on the Balkan Circle has been particularly worthwhile, and I want to thank all attendees and presenters of this blossoming initiative. Thanks to Vlad Beronja and Chelsi West Ouheri, whose work, feedback, and presence in the Slavic Department have been amazing.

I appreciate so many colleagues I have connected with at ASEEES, other conferences, and through varied scholarly interactions over the past few years. Thank you all for your comments, feedback, and encouragement. This includes, but is not limited to, Robert Nemes, Andrew Behrendt, Alison Orton, Wendy Bracewell, Hillel Kieval, and Larry Wolff. A big thank you to Lyubomir Pozharliev (and Stefan Rohdewald!), who invited me to Giessen, Germany, to give a talk and a master class on food history. I am always gratified when professional interactions lead to friendships, and I hope to continue our collaborations in the coming years.

Speaking of colleagues who are now good friends, I have to thank Paulina Bren, who has always been such a good friend and helped move forward my thinking about consumption under socialism that took form during our coediting of *Communism Unwrapped*. She continues to be a source of friendship and encouragement. I also want to thank Ali Igmen and Kate Brown for always being there for me as friends and colleagues. I met all three of you—Paulina, Kate, and Ali—back at the University of Washington–Seattle in the 1990s; you have all been pillars of support. As to newer friends, thank you Choi for everything! I so enjoyed giving a talk and meeting your (and Ali's) students in LA and enjoying a meal in your lovely (backyard-to-table) garden. I also want to call out the incredible Maša Kolanović, who never stops inspiring me with her creative thinking and boundless intellectual energy. I can't wait for our next trip to Dugi Otok, or for you to come back to Austin.

I can't thank enough all the people who have helped me on the ground in Bulgaria. In addition to Bulgarian colleagues listed above, I want to thank my friend and colleague Mariana Stamova, whom I have known now for over twenty years. For at least the last ten she has been helping me navigate the changing regimes and rules of the Bulgarian state archives. I am extremely grateful to her, but also to the dedicated archivists and librarians in Bulgaria who make my work possible—especially Mikhail Gruev and Iliana Paskova of the Central State Archive. I also want to thank the archivists at the regional archives in Plovdiv, Haskovo, and Blagoevgrad. And finally, a thanks and a goodbye to one of my favorite Bulgarian historians, the late Kostadin Grozev—I will miss our long dinners and laughter at the Bizhu restaurant in Sofia.

Finally, I want to thank my family, my kids—Sophie, Bella, and Dean—who grew up to accept and support the fact that their mom travels a lot and devotes countless hours to researching and writing about a distant time and place. Last, but not least, thank you Jeff for following me to the Balkans (more than once) and appreciating the places and people that I so love. Thank you for our countless conversations over food and wine.

Chapter 5 is a version of an article published in *Contemporary European History* 29, no. 4 (2020): 416–30, under the title "Drinking to the Future: Wine in Communist Bulgaria." In addition, much of the material from chapter 1 was used in the edited book chapter "Hungry for Revolution: Women, Food and the Bulgarian Left, 1917–1923," in *Wider Arc of Revolution: The Global Impact of 1917*, coedited by Mary Neuburger, Choi Chatterjee, Steven Sabol, and Steven Marks, Slavica Press, 2019.

INGREDIENTS OF CHANGE

Introduction

I came to love the Balkans first and foremost via food.[1] On my first visit to the region in 1993, I was captivated by the sheer perfection of a Bulgarian tomato, the gamey richness of local yogurt, the delicate piquancy of a *kebabche* (grilled meat patty), and the lush earthiness of Bulgarian red wine. Bulgarian food—its color, flavor, and texture—was one of my most immediate and intimate daily points of contact with the landscapes, locales, peoples, and even animals of the Balkans. Like all travelers, past and present, I had an overwhelming impulse to evaluate my surroundings through food—with the anxieties, pleasures, and revelations that inevitably abound. Food is one of the most elemental and most intimate means of exploring the "other," embedded in our quest for exotic or authentic "foreign" experiences.[2] But this quest is also ultimately about discovering our individual and collective selves, about divining not just what we need to survive or thrive, but also what we love, hate, and aspire to.

Balkan cuisines as a palatable example of the entangled and contested cultures of the region have long piqued my culinary and intellectual curiosity. In Eastern Europe, the culinary topography has been shaped in large part by imperial legacies. Food geographies reflect the fluctuating borders of the Russian, Habsburg, and Ottoman Empires, which ruled this vast region until their dissolution in the nineteenth and twentieth centuries. Modern elites eventually carved out their own national cuisines, laying claim to common food and drink traditions as denoting *national* distinction and authenticity. Such claims belie the melting pot of shared foods that, in the case of the Balkans, stretches from Sarajevo to Sofia to Salonika, from Tirana to Timisoara—and beyond.[3] For a large swath of the Balkans, connective culinary threads are woven together as variations of a regional cuisine. This is a clear reflection of shared climate and historical experience, most notably the period of Ottoman rule from the fourteenth to the nineteenth centuries and communist rule from 1944 to 1989. And yet such common Balkan dishes as *banitsa/burek/börek* (puff pastry with meat, cheese, or other fillings) and ingredients like feta cheese and yogurt are commonly claimed by individual nations as integral parts of their own distinct national cuisines or foodways.[4] Why this insistence on national labels?

In recent years, food and drink have captured the imagination of a global array of scholars and publics hungry for new ways to eat, arrange, and celebrate food, and to ponder its many forms and meanings. As a rapidly expanding scholarship on food has revealed, national cuisines, like national identities, are the product of a modern political and cultural collective imaginary.[5] Such works explore how national identities are accompanied or, in some cases, followed by the emergence and evolution of national cuisines. Like nations, such cuisines are a product of local, regional, and even global encounters with "other" cultures, ideas, plants, animals, and products.[6] This growing body of work is most prolific in the case of the United States and Western Europe, with Asia and Latin America not far behind.[7] Russian food studies have also come a long way in recent years, with a number of in-depth works on the varied fates of food and drink before, under, and after communism.[8] Eastern Europe, in contrast, has been slower in coming to the table of food studies, although interest in the region's foodways is growing—particularly in connection with the larger field of consumption under and after communism.[9] But more and more scholars in and outside the region are turning to the rich subject of food.[10] This book attempts to build on a rich new body of work on the history of Bulgarian food.[11] It seeks to flesh out the *longue durée* history not just of cuisine, but of what underpinned the modern, even revolutionary, transformation of foodways in Bulgaria—that is, changes in what, how, and why people ate what they ate.

Part of this story concerns cuisine, the gathered recipes—themselves assemblages of ingredients and techniques—that are curated into a notionally, nationally coherent whole. That is, how and when did Bulgarian cuisine emerge as a particular ordering of ingredients? What were the influences, not just culinary, but also political and scientific? One of its main influences was Ottoman cuisine, which was itself an amalgam of inspirations—Central Asian, Persian, Anatolian, Greco-Roman, Levantine, and Balkan. It was shaped by the multitude of peoples, foodstuffs, and techniques that originated in or passed through the empire's near and far-flung provinces, and especially Istanbul, the hub of culture and exchange.[12] Through Ottoman trade centers, new kinds of foods from the Far East and the New World—spices, plants, and more—spread across the Balkans. The Orthodox Christian Slavs of the region (today's Bulgarians)[13] both discovered new foods and were agents in their diffusion in the eighteenth and nineteenth centuries. As a core Ottoman province from the fifteenth to the nineteenth centuries, the region that is now Bulgaria was an important food provider, boasting fertile valleys, biodiverse uplands, and a long growing season tempered by the warm Black and Aegean Seas. Ottoman Bulgaria was dotted with mostly small but productive subsistence farms, but it was also home to larger rice and wheat plantations, fertile gardens, and herds of livestock, primarily sheep and goats. Bulgarians were an important part of overland trade networks, which brought animals on the hoof as well as animal and plant products to Istanbul and other major cities, in the Ottoman as well as the Habsburg and Russian Empires. Bulgarians were also important itinerant market gardeners in cities across these three empires, growing fresh produce and selling it to urban dwellers. In this way they impacted local foodways, even as they also absorbed new ways of producing and consuming food and drink.

Although the focus here is on the region that is today Bulgaria, the process of transformation was inseparable from regional and global processes—the circulation of powerful food narratives, along with peoples, goods, technologies, and political paradigms. In particular, along with Ottoman practices, Central European and Russian influences were gradually incorporated into Bulgarian and Balkan foodways. This intensified after Bulgaria's "liberation" from the Ottomans in 1878, amid shifting ties to Russia, the Habsburg Empire, and Germany. Bulgarian foodways were altered in the most revolutionary ways in the course of the twentieth century, given new kinds of cultural anxieties, social mobility, and evolving tastes. These changes happened in the shadow of key events such as the devastation of the Balkan Wars and two world wars. Even more critical were the transformative effects of communist rule from 1944 to 1989. During this period, Bulgarians continually appropriated global food narratives (scientific, religious, ethical) and techniques (culinary, managerial, technological, commercial), even as those narratives and techniques were domesticated, adapted, and/or resisted.

The "West" in many respects was the most critical interlocutor or foil in this process, a model or counter-model, a source of inspiration or a malicious force to be resisted.[14] In modern Bulgarian history, as in Balkan history more generally, the desire to be accepted as "European" has always been in competition with ambivalence toward the homogenizing process of modernity.[15] A variation on this theme, of course, can be found across the region, and in non-Western societies around the globe.[16] One of the subcurrents of this book is an exploration of the ways in which the question "Should we *be* like the West?" became linked to the question "Should we *eat* like the West?" This tied into the larger calculus of taste, nutrition, etiquette, and other *modern* values, beliefs, and practices. The answer was never simple, as the West continually inspired emulation or spawned counter-impulses. In either case, food practices from the Ottoman and pre-Ottoman eras offered authenticity and a kind of culinary pedigree, as well as models of refined or restrained consumption. Both the past and present provided a pantry of ingredients for looking forward—recipes for the future, which could be reassembled and consumed in limitless ways.

This book is most concerned with how modern changes in Bulgarian food production, consumption, and exchange were driven by an evolving modern bioimaginary—that is to say, how individuals, collectives, and agents of the state imagined their present and future through food and drink, its ingredients, biochemistry, sensory properties, and possibilities.[17] I am also interested in the ways such imagining—religious, secular, and scientific—translated into biopolitics. Here I use "biopolitics" to describe the ways in which a variety of actors, especially state actors, intervened to make decisions about how bodies should be nourished, disciplined, regulated, and mobilized.[18] Through a diverse range of sources, from state, local, and international organizations, to archives, trade, scientific, and popular journals, cookbooks, novels, and travelogues, I trace the ways in which global food narratives were appropriated, refracted, and domesticated in the Bulgarian context. The story unfolds through focused chapters on what I see as key ingredients of change—namely, bread, meat, milk (and yogurt), fruits and vegetables (tomatoes and peppers), and wine. I uncovered the intriguing past of each of these distinct constituent parts of the Bulgarian diet as I traced their particular "social lives" and mythologies.[19] I feel that this approach provides the best recipe for success in understanding Bulgarian food, as both intimately local and inextricably tied to regional and global food systems and narratives. Researching the history of Bulgarian food over a century of change opens a range of possibilities, but also a veritable can of worms. With that in mind, I offer this book not as a comprehensive or complete history, but as a starting point on the way to a relatively new and promising field for Balkan studies. I hope to insert this Balkan "periphery" into the larger story of global food, while also re- or de-centering that story.

Ingredients of Change

This book covers more than a hundred years, but its core questions and arguments revolve around the post–World War II period, when Bulgaria was under the one-party rule of a communist regime. This forty-five-year period (1944–89) arguably brought equal or even greater transformation than five hundred years of Ottoman rule. It witnessed momentous changes in the political, social, economic, and cultural realms, including scientific research and education, agricultural practices, food processing, consumption patterns, and trade.

It was under socialism that the imperative to "catch up" to the West set in motion a transformative new bioimaginary and resultant biopolitics in which the fortifying of bodies was integral to the drive for progress. In part this was a reaction to nutritional narratives and policies coming out of the Western scientific community and intergovernmental organizations such as the League of Nations and the United Nations. In particular, twentieth-century Bulgarian statesmen and intellectuals became preoccupied with the notion that fully nourished bodies were not just a reflection of development, but a requirement for progress. As elsewhere in the Eastern Bloc, "full nutrition" fueled the human machine that would build socialism. A new reliance on the scientific management of food production, consumption, and exchange meant counting everything, from pigs, to tons of wheat, carbohydrates, protein, and vitamins. This calculus framed the entire Eastern Bloc's alternative food system, through the Council for Mutual Economic Assistance (Comecon), formed in 1949. This integrative body planned economic cooperation and specialization across the Bloc, including nutritional targets, which attempted to balance Bloc-wide production with regard to protein, carbohydrates, and vitamins. Within Bulgaria, a range of sources and state actors worked to fuel a popular belief in food science, with carefully calculated formulas of bodily requirements for calories, protein, vitamins, minerals, carbohydrates, and fat. These formulas were embedded in the new state-supported conviction that a specific combination of nutrients was needed to build and fortify industrious bodies. Postwar food science evolved in direct accordance with Soviet, as well as global, theories and norms, but it was also considerably domesticated. In many cases, food science consecrated time-honored assumptions about health and food, becoming the new religion, with its own prophets, priesthood, and revelations. Like religion, science required *belief*.

With this changing bioimaginary as a backdrop, this book paints a picture of what I see as the momentous, and in many respects successful, transformation of Bulgarian foodways during the Cold War. To be clear, I am not trying to glorify communism as a system or to ignore its obvious costs in human lives, freedom, and dignity. One cannot overlook the brutality of the immediate postwar imposition of power, the people's courts, show trials, and collectivization, which required a brutal crushing of resistance, especially from 1944 to 1951. But a large segment of the population also supported and benefited, individually and collectively, from various aspects of the socialist system. Bulgarians, like others in the region, continually reassess communism and its legacy. Nostalgia and condemnation compete with more nuanced histories of this recent past.[20] My focus here is not on the system's flaws or failures, which could fill countless books in their own right, but on the real and rather remarkable, albeit imperfect, food revolution that occurred in Bulgaria from 1944 to 1989.

It is worth remembering that the Soviet Union was born of hunger; the Russian Revolution was about bread.[21] Across Europe, the final years of World War I brought a hurricane of hunger, which leveled the continent and made radical visions of the future, like communism and fascism, possible. The Soviets used food, including bread, as a tool for establishing and maintaining power, especially during the forced famine in Ukraine.[22] Collectivization was catastrophic in the short term, as was the Stalinist hold on biology in the form of Lysenkoism, but the Soviets also made great strides in food processing and agricultural science.[23] The Soviet Union became a kind of mother ship for the interwar Bulgarian left, which was eventually driven underground and abroad by the pro-German right. But leftist ideas would continue to percolate at home and among émigrés in the USSR and elsewhere. The dreams of the pro-Soviet Bulgarian left were realized as World War II ended and the Red Army occupied Bulgaria. In the period that followed, Bulgaria and the rest of the new Eastern Bloc underwent a radical transformation, which included the dispossession of former elites as well as a brutal collectivization process.[24] Collectivization clearly had political as well as economic objectives, but feeding the masses was always a key concern, as well as a foundation for power.

Contrary to popular beliefs about this period, one can speak of a generalized abundance of food in postwar Bulgaria. While never devoid of problems and deficits, a successful food system emerged.[25] In terms of the standard of living, the socialist system, with its radical reallocation of resources and mechanization of production, brought significant improvement to the bulk of Bulgaria's largely rural population.[26] As far as the state was concerned, properly fortified bodies were needed to build socialism, just as well-fed citizens were needed to manufacture political consent. An integral part of building this consensus was economic change, including lowering the cost of foodstuffs—a cornerstone of the socialist "social contract." As many scholars have argued, the rising standard of living contributed to a new kind of stability, a normality in everyday life under communism, which spread across the region by the 1960s.[27] Indeed, food consumption increased dramatically in Bulgaria as urbanization and a rapid expansion of higher education and job opportunities gave rise to a new communist middle class.[28]

By the 1960s, the communist state catered to, and sometimes even encouraged, a new culture of food, in which nutritional objectives were coupled with notions of taste and connoisseurship. Food culture was a constituent part of the process of "civilizing" newly urbanized populations, a kind of uplift through modern modes of eating and drinking such as nutritional standards, etiquette, table arrangements, cooking techniques, and public eating venues.[29] The 1960s and 1970s witnessed a massive increase in the number of restaurants and other food venues in Bulgarian cities as well as tourist locales. Though excessive bourgeois consumption remained subject to communist-era critique, in practice there was a kind of embourgeoisement of the Bulgarian citizen consumer, in which food culture played an important part. Women had a special role as apostles of the new culture, charged with managing, fortifying, and civilizing the bodies of their families. In spite of their supposed "liberation" under communism, women made a critical contribution in both feeding their families and implementing the new food culture. Food was a key ingredient of what scholars have termed the socialist "good life," namely the promise of material security, abundance, and everyday pleasures for all.[30] If the communist leadership needed evidence that the promised Marxist utopia was just over the horizon, food was a slice or glass of utopia in the here and now. It was simultaneously a herald of (if not a replacement for) the illusive dreamworld, the bright future promised by Marxism.

Interestingly, the new socialist food culture also sought to establish, or imagine, a European cultural-*cum*-culinary pedigree. This was about more than "civilizing" the populace; it was a matter of staking a claim to a culinary heritage for Bulgaria's "European" foods—like bread, cheese, and wine. In a society obsessed with the future, it was important to reorder the past in a way that obscured Ottoman influences. An array of official sources from the 1970s and 1980s, for example, cite the ancient Thracians—some of the continent's earliest producers of cheese, wine, and bread—as having laid the foundation for Bulgarian food culture. For some this undergirded a Bacchanalian embrace of the pleasures of food and wine, which accorded with the model of the socialist good life. And yet, such formulations competed with notions of social restraint, also scaffolded by Bulgarian tradition—ancient, medieval, and modern. Visions of consumer restraint also came from official sources, which drew upon traditions of ethical or spiritual rather than material success. Ironically, as the period wore on, the past was needed more than ever to scaffold a bioimaginary that was focused on a bright future.

Amid these competing late socialist narratives, shortages of food and other goods began to deepen an already brewing crisis of legitimacy both in Bulgaria and across the Bloc. In many respects, the Bloc was a victim of its own success, as the relative abundance in the 1960s and 1970s had led to a rise in expectations that made the well-known shortages of the 1980s seem even more severe. If a variety of foods were in short supply, people were still well fed, just largely inconvenienced in a system that required connections, queuing, black-market trading, home canning, and other gyrations to put food on the table. Bulgarians, like other East Europeans, "made do," procuring food through personal connections, black-market networks, hoarding, and do-it-yourself cultivation and canning.[31] Even as agriculture became mechanized, small plots remained an important part of the supply chain. If this was good in terms of a sustainable agricultural system requiring fewer chemicals, for many it was also a continuing disappointment—and perhaps the beginning of the end—especially as Bulgarians, like the rest of the Bloc, began to compare their lot to the abundance of Western capitalism, which had come into its own by the 1980s.

This propensity to compare and compete, as noted above, was by no means new in the 1980s. The Cold War was a period of systemic competition that drove change on both sides of the Iron Curtain. A critical part of that competition took place along the road from farm to table. The United States established its place as a world power largely because of its ability to feed the European continent after both world wars.[32] Once postwar hunger was under control, East and West endeavored to prove the superiority of their own system in providing a more "complete" diet. Cold War development narratives were tied to the notion that the largely grain-based diets of non-Western societies needed to be replaced with a diet featuring ample supplies of animal protein and the vitamins found in meat, milk, and produce. Under the influence of the international scientific community, such narratives had circulated through the League of Nations, and then the newly formed United Nations and its specialized agencies, the Food and Agriculture Organization and the World Health Organization. While these organizations were largely focused on the "developing world," their overarching nutritional paradigms—in particular the focus on protein and vitamins—were largely appropriated by the Eastern Bloc. Any hope of catching up to or overtaking the West through modernization and development seemed impossible without protein- and vitamin-fortified bodies. By the end of the Cold War, protein and vitamin consumption was off the charts behind the Iron Curtain as diets newly rich in meat, milk, and produce became the order of the day. But the very deliberate Bloc-wide shift away from grain toward a diet rich in meat, dairy, vegetables and fruit had also meant a costly reorientation from cultivating grain for human consumption to producing it for fodder, as well as toward a richer variety of produce, and "luxuries" like wine, spirits, and tobacco.[33] As early as the 1970s, shortages of wheat created the need for mass imports from the West, causing debt to soar across the Bloc. It was not *just* food, of course; but food became a critical factor in the debt cycle and crumbling of legitimacy as Bloc economies imploded in the late 1970s and 1980s, which led to eventual systemic collapse.

If the United States, in contrast, weathered the shocks of the 1970s and thrived in the 1980s with fully stocked supermarkets, we also have to question the notion of its unmitigated "success"—especially in terms of foodways. One of the subtexts of this book is a pointed questioning of the prevalent notion of a triumphant American Cold War food system in relation to the Eastern Bloc's "failure" to provide. In terms of food production and supply, there is no doubt that the United States was able to outproduce not just the Eastern Bloc, but the rest of the world, including Western Europe, whose populations it played a part in feeding. Strictly in terms of numbers and output, the successes are extraordinary. To be clear, this would not have been possible within a purely free-market framework. That is to say, agriculture in the United States, as everywhere, has always been subject to government direction and subsidy, more than any other sector of the economy. More pointedly, multiple studies of American food and its role in the global food supply tell a rather dark story, with "successes" coming at a high cost. The US Cold War food system, with its mono-cropping and encouragement of large-scale production, had a destructive effect on small-scale farming, and on "heritage breeds" of plants and animals.[34] As a consequence—and as most of us can taste for ourselves—American industrialized food was stripped of flavor, texture, and nutrients in the drive to mechanize. One of the costs of American (and American-modeled) agricultural development has been environmental devastation—the pollution and waste of water, the destruction of soil, and the loss of biodiversity. Indeed, industrialized agriculture has done more to destroy the environment than any other human activity, and its current practices will not be sustainable in the future. This pollution has been mirrored in the effects on human bodies, with a third of the United States and a fourth of the world (under Americanized regimes of eating) now obese.[35] Chronic disease related to overconsumption of certain foods—processed and otherwise—is by no means a winning prospect. And it is not just the rich who are in this category. The fact that food production and exchange are so inexorably tied to large profits for corporations perpetuates this model; a food system inundated with cheap processed foods means that large numbers of people are sick, whether they are "stuffed" or "starved."[36] All of this is just for context, to point out that assumptions about "Western" success in relation to the Eastern Bloc should be tempered by the reality of the unsustainable and arguably broken food system that came into its own during the Cold War period.

I can't help but recall a particular moment in the food memoir of Anya Von Bremzen, who emigrated from the Soviet Union to the United States as a child in the 1970s. She was bitterly disappointed to discover the "dark truth" about American food: it was "not exactly delicious." The "rubberized" bread made her pine for the "dark burnished loafs" of Moscow.[37] Before the US foodie scene exploded in the 2000s, "real" food was not always easy to come by, at least in American cities. It was part of a fringe countercultural movement that only later crept into the mainstream.[38] Even now, one could argue, "slow food" is still in many respects affordable only for a select elite, while large numbers of Americans live in "food deserts" and eat highly unhealthy processed foods. In Europe the situation is far less dire, but it is a sustained challenge to Americanization that has allowed the food system to diverge in many respects from the postwar American model.[39]

Putting American failures aside, it still makes little sense to compare Bulgaria's Cold War "progress" in terms of food to the "West," let alone the United States. Bulgaria's level of development was nowhere near that of the United States before communism, and it certainly has not been since. Yet communist successes and failures have generally been cast in comparison to distant and unattainable ideals. For that reason, the United States and, to a lesser extent, Western Europe enter this story in places as players, and for context rather than overt or systematic comparison. The West is an almost constant frame of reference in this period of transformation, as a potential model for change, but also as a cautionary tale. As I will argue, Cold War boundaries created a kind of alternative Eastern Bloc food system, albeit one that was not hermetically sealed. Within this system, Bulgarians' deeply embedded yearnings and resistance in relation to the West came into play as they adopted, but also adapted, Western technologies, while continually encouraging and relying on "traditional" or newly introduced food preservation techniques. Ultimately these were more sustainable methods of food production and consumption, although Bulgarian political and economic systems proved unsustainable.

One way to assess the communist past is to look at what followed it, and I provide at least a taste of the transition in this book. In the immediate postcommunist period, there was economic chaos, which brought a massive drop in calorie intake and what seems to be a permanent disruption of the local food system, now subject to the challenges of globalization.[40] While it is true that ultimately the postcommunist period brought choice, with an infinitely greater variety of goods (and restaurant options), for many this new cornucopia is out of reach. In fact, nostalgia for the communist period in Bulgaria, which has ebbed and flowed since 1989, is at a high point in the current fiscal crisis. The lack of spending power today, particularly for the elderly, unemployed, and rural populations, has profound implications for the ways in which people remember the communist past. Consider an oft-cited Bulgarian joke:

> A woman sits bolt upright in the middle of the night in a panic. She jumps out of bed and rushes to the bathroom to look in the medicine cabinet. Then, she runs into the kitchen and opens the refrigerator. Finally, she dashes to the window and looks out onto the street. Relieved, she returns to the bedroom. Her husband asks her, "What's wrong with you?" "I had a terrible nightmare," she says. "I dreamt that we could still afford to buy medicine, that the refrigerator was absolutely full, and that the streets were safe and clean." "How is that a nightmare?" the husband asks. The woman shakes her head, "I thought the communists were back in power."[41]

When Bulgarians are feeling nostalgic, they generally recall the more abundant 1960s and 1970s, when shelves were relatively full and, more importantly, salaries and pensions were sufficient to provide a higher standard of living. They also remember the security, in terms both of low food prices and of the affordability of education and health care. What is more, under communism as many people remember it, the crime rate was low, there was public support for Bulgarian high culture and scholarship, and a sense of order prevailed.

This is in stark contrast to the seeming chaos of post-communism, the perception that everything is for sale and nothing is cheap (including food). Capitalism and democracy have brought opportunities and new freedoms, but also significant costs. Rampant corruption and almost continuous fiscal crisis have been coupled with brain drain—whole generations of the best and the brightest leaving Bulgaria for better, distant opportunities. All of this has driven a deepening sense of a loss of coherence and order, and a total lack of trust in the current system.[42] When it comes to food, the issue is not just price or quantity, it is substance and quality. In the postcommunist present, stringent EU regulations and competition have severely altered the food system, shutting down "unhygienic" local operations and bringing in multinational corporations. A plethora of new processed goods—and fast food, including fifteen McDonald's—have flooded the market. In such an environment, memories of "real" food often eclipse some of the less savory realities of the communist past. But as noted above, food under communism was less a product of time-honored tradition and more a work in progress, even a revolution in the way people ate. The chapters that follow explore the ingredients of this change—looking before, during, and after communism to unpack the complex farm-to-table processes through which a stable Bulgarian cuisine emerged.

 CHAPTER 1

By Bread Alone?

Hunger, Abundance, and the Politics of Grain

In May of 1918, hundreds of mostly female workers in Plovdiv, Bulgaria, took to the streets with signs reading "We want bread," "We want our men," and "We want an end to war."[1] In the final years of World War I, severe grain shortages left large swaths of Europe hungry, even starving. In Bulgaria, bread rations were meager, unavailable, or inedible, as bread was made with sawdust and other fillers. During the May riots, angry women converged in front of the municipal food-provisioning center in Plovdiv yelling for bread and throwing rocks at the windows. Unable to find either provisions or answers, the group proceeded to the mayor's house, where the desperate women cried out for bread until the local militia arrived and fired on the crowd, killing at least one female protester. The women were irate. They threw more rocks and broke into the mayor's cellar, distributing whatever they could find—flour, cheese, soap, chicken, and other stockpiled supplies—to the gathered throng.[2] This was just one of the hundreds of so-called women's revolts that exploded across Bulgaria in the spring of 1918, mobilizing thousands of hungry women to protest home-front hunger exacerbated by food requisitions, speculation, hoarding, and exports to Germany.[3] Across the country, urban women converged on storehouses held by municipal administrators and spontaneously redistributed goods to the assembled crowds. Rural women resisted continued requisitions of food or forcibly took back animals, grain, and produce from collection stations.[4] Many met with arrest, injury, and even death at the hands of local police or militia, which spurred even greater discontent.[5]

More than any other category of food, grains—and more specifically bread—have been deeply entwined with power, politics, and revolution in modern world history. In a sense, grain was perhaps even more foundational to modern state power than weaponry: it allowed polities to feed cities and armies and thus secure sovereignty in the ancient, medieval, and early modern worlds. The ability to cultivate, amass, procure, store, and transport grain was the basis for population growth and urban development, and essential as well for conquest, which required feeding armies on the move.[6] Providing the population with grains and bread remained a primary requirement of twentieth-century political legitimacy, even as the extension of food supply chains made the food system increasingly vulnerable to interruption, if not revolution. World War I brought catastrophic disruptions across Europe, exacerbating hunger and desperation on war fronts and home fronts alike. In the shadow of mass hunger, women's revolts—which largely took the form of food riots—played an important role in the wave of revolutions that shook Europe in the final days of the war and its aftermath.[7] In Russia, bread riots served as a critical catalyst in bringing the tsarist system to an end in 1917, while requisitioning, controlling, and ultimately providing bread and grain were central to Bolshevik and Soviet success in establishing and maintaining power.[8] Later, bread and grains would play a major role in the Cold War, in terms of both scaffolding domestic legitimacy and framing geopolitical competition. This rivalry, however, which was founded on the ability to provide bread, moved ultimately to the necessity of moving beyond it, to providing much more.

This chapter explores the place of bread in modern Bulgarian history, which was deeply rooted in the soils of the eastern Balkans, but also attuned to the winds of regional and global change. Bread was a central feature of the Bulgarian diet, but it played an important role in the changing national bioimaginary. Bread's place in the Bulgarian diet changed dramatically in the course of the modern period in the shadow of new ideas, regimes, and possibilities. Bread (or the lack thereof) had clear political revolutionary potential; the mass hunger that fueled the Bulgarian riots of 1918 was a severe shock in a country that had long been able to feed itself and even export grain. But the revolution in bread was most pronounced after World War II, when the Bulgarian Communist Party established its authority by promising and supplying bread as it established control over the food system.

In some respects, the Cold War was fought not on the battlefields, but in the wheat fields and on the tables of the twentieth century. Bread found its place in the Bulgarian communists' vision for forging a new society through the efforts to create a "new man," which required both a nutritional and a civilizational transition for the Bulgarian masses. The new man was to be formed in part through new attention to biochemistry, namely through "rational consumption" of food—that is, a science-based balanced and nutritional diet. But the communist goal was also to bring to the new socialist man the "good life," which went beyond satisfying basic needs—like bread—to include a cornucopia of pleasurable, modern consumer practices. This meant not just a steady supply of bread, but an abundance of newly leavened and whitened breads and pastries (which in the past had been reserved for holidays). It also meant a greater proportion of meats, fruits and vegetables, dairy, and other foods, both to balance the traditional bread-based diet nutritionally and to infuse it with pleasure and richness.

Needless to say, "rational consumption" and the socialist "good life" were at times in conflict, as the story of Bulgarian bread will amply illustrate. For Bulgaria, as for the rest of the Eastern Bloc, utilitarian notions of nutritional eating were at times undermined by efforts to provide foods—like mass-produced leavened white bread and pastries—that were seen as harbingers of status, success, and progress. By the late socialist period, the problem was not a lack of bread. Indeed, by the 1960s, Bulgarian and Bloc planners and nutritionists were concerned that citizens were consuming *too much* bread in relation to other foods such as protein-rich meat and milk. The latter were needed not only to build and fuel socialist production, but also to help the region "catch up" to the West. The population, however, was not ready to curtail its consumption of bread, which was supplemented and not supplanted by other changes in the diet. At the same time, bread in its newer, whiter, fluffier varieties was appealing to the populace, but it was less nutritious and so went against the grain, so to speak, of "rational" eating practices. In some respects, the fatal flaw of the communist system was that bread, even in its newly modern form, was not enough for the promised good life. More importantly, it was a hindrance to communist-era aspirations of going beyond—or even against—bread and grain as the center of the Bulgarian diet.

The Essential Loaf

It is no surprise that grains have had such momentous political and cultural significance in global history. That is especially the case for wheat, which, as historian Felipe Fernández-Armesto points out, "has spread over the world more drastically, invaded more habitats, multiplied faster, and evolved more rapidly without extinction than any known organism."[9] Grain played an important role in human settlement, providing sustenance to ancient civilizations and early modern empires, including the vast Ottoman Empire. The Bulgarian lands were among the core provinces of the empire, some of the earliest Balkan provinces to be conquered in the late fourteenth century. The rich alluvial valleys of this region—particularly Thrace and Dobrudja—provided grain to Ottoman cities and armies, feeding humans and animals alike. The Black Sea and the navigable Danube River became essential conduits for transporting Balkan grain to Istanbul and other Ottoman port cities. Grain was a key source of wealth for the Ottoman state, which taxed the peasantry in kind, most often in grain—not just wheat, but also rice, barley, millet, and eventually corn.[10] From the imperial capital to the smallest provincial villages, an enormous variety of breads joined boiled grains as a staple of the Ottoman diet for everyone from elites to the most modest peasants.

The reverence for bread in the Bulgarian settled lands dates back to well before the Ottoman conquest. Bread had been consumed in the region since ancient Thracian times, as well as by sixth-century Proto-Bulgarians. In the medieval period, unleavened bread (or in some regions boiled grains) made from wheat, rye, oats, or millet was the primary source of sustenance for the bulk of the population. Bulgarians and other Slavs of Eurasia quite literally had a "cult" of bread woven into their systems of folk belief, shaped by both pagan and Christian tradition.[11] Bread in a sense *was* God, the presumed product of fertility prayers and (when blessed) the body of Christ in Christian tradition.[12] Bread was consumed every day by Balkan peasants and urban dwellers alike, with accompaniments varying by the season, but almost always including salt, onion or garlic, and wine.[13] As discussed in depth in later chapters, meat and dairy products were considered something of a luxury for the peasants and urban poor of the region; they were regulated by the Orthodox calendar, which included 182 meatless (and vegan) fast days a year. Unleavened *piti* (flatbreads), made with a variety of whole grains, dominated the rural diet well into the twentieth century, often sprinkled with salt or some variety of spice, such as hot red pepper, savory, or thyme.[14] But grains also were used in a huge range of concoctions, including various fried, roasted, and leavened baked breads, as well as porridges, soups, and even dips made from leftover dried bread. As Spas Raikin, who was born and raised in interwar Zelenikovo (near Plovdiv), describes,

Our breakfast more often than not consisted of leftovers from the day before, if there were any. Most of the time it was *popara*—dry bread soaked in boiling water, sweetened by one or two spoons of *petmez* [a sweet syrup of beets]. . . . Sometimes the *popara* was replaced by *trahana*—little crumbs of yeast from the *noshtvi* [bowls used for kneading bread], mixed with a little flour, boiled in water and served as a sour soup. . . . Another delicacy was the *ushmer*. It was prepared from flour, fried in a pan, and then some water would be added and boiled to desired thickness.[15]

His account goes on for pages, with descriptions of flour and, less commonly, unmilled grains being fried, baked, and boiled into a myriad of forms, then eaten with a range of accompaniments. These different ways of using grains provided variety in a diet largely dictated by poverty.

In general, bread—rural or urban, leavened or not—was baked in communal ovens, which in many cases were run by guilds, though flatbreads (*piti* or *pogachi*) could be cooked at home on a grill or in and among the coals of outdoor fires or wood-burning ovens.[16] By the end of the nineteenth century, a growing number of urban bakeries regularly produced and sold leavened loaves made with refined white flour. In this period, white bread became associated with urban wealth, and by association Europeanization and modernity.[17] And yet, as Bulgarians urbanized and a new elite formed, culinary taste was also closely aligned with the spice and flair of Ottoman cuisine. This included a tradition of more elaborate pastries, generally sold in the coffeehouse (*kafene*) or by street vendors. The most enduring of these was the *banitsa*, which could be sweet or savory, but was most often filled with crumbled feta-style cheese and butter. Commonly called *biurek* in Turkey and elsewhere in the Balkans, *banitsa* remains among the most popular savory pastries across the region. It is one of many Ottoman legacy dishes that have been appropriated by Balkan national cuisines.[18]

Ottoman cuisine grew out of an amalgam of Near Eastern (especially Persian) and Central Asian food practices and ingredients. It was as diverse as the three continents it straddled and their patchwork of peoples who—along with a myriad of foreign nationals—settled in or passed through the imperial capital.[19] In Ottoman Istanbul, the diversity of available foods and culinary influences was staggering, echoed on a smaller scale in the provincial capitals. In the course of the nineteenth century, new elites among the primarily rural Christian Slavic population began to urbanize and also identify as "Bulgarian." They formed a kind of archipelago of urban colonies in the major cities of the empire, as well as in Vienna, Odessa, and elsewhere in Europe. The vast majority of Bulgarians—by some estimates as many as fifty thousand—lived, worked, traded, studied, and found sustenance in the imperial capital, which was alive with bustling spice and food markets and street vendors. It was there that Petko Slaveikov, who became one of the most famous Bulgarian literary figures of this era, wrote the first cookbook in the Bulgarian language in 1870.

Slaveikov's *Gotvarska kniga* (Cookbook) offered recipes and cooking techniques from the Ottoman capital to Bulgarian readers everywhere.[20] Interestingly, flatbreads were virtually absent from its pages, as was *popara* or any type of dried bread porridge. Only two actual "bread" recipes were featured, both yeasted: "good and simple bread" (in which boiled rice was added to a simple loaf of unspecified "flour" and yeast) and "bread from apples" (bread with chopped apples added).[21] The cookbook represents a clear push for a more urban and elevated bread-making culture, which extended to other grain-based dishes, such as an elaborate array of pilafs made with spices, eggplant, tomatoes, lamb, chicken, or mussels. *Gotvarska kniga* also provided recipes for all manner of *banitsa*, made with eggs, butter, sugar, and meat, as well as baklava.[22] Interestingly, Slaveikov's cookbook reflected neither the common Bulgarian peasant diet nor an effort to define a distinct Bulgarian national cuisine. Instead it offered a set of culinary practices as a model for the new Bulgarian elite to elevate and modernize their diet, which reflected their peasant past.[23] Bread, which was already on the daily menu, did not play an important role in such a project. Instead there was a focus on meats and dairy, sweets, pilafs, spices, and baked goods that were less common in the Bulgarian diet—for reasons of price, access, and preparation time in rural settings.

There was something portentous in the sidelining of bread in this aspirational version of the Bulgarian diet. Subsequent political revolutions, such as the women's revolts of 1918, may have been undergirded by demands for bread, but for many revolutionaries the goal was to move beyond *just* bread, or living by bread alone. By the interwar and postwar periods, gastronomic ambitions were inspired by the protein-packed diets of the "civilized" West. But in the nineteenth century it was Ottoman elites, pashas and richer urban dwellers, whose dietary cornucopia was the primary model for emulation. Interestingly, the West itself had long drawn inspiration from the East in terms of spice and culinary variety as well as other consumer practices. The new Bulgarian merchant elites in this period looked to readily available Ottoman models of consumer fulfillment, and cuisine was no exception. At the same time, for a segment of newly educated Bulgarian elites,

FIGURE 1.1. Women preparing flatbread in the village of Berkovitsa, Bulgaria, circa 1930s. Library of Congress, Prints and Photographs Division, Washington, DC.

Ottoman rule seemed to pose an impermeable barrier to full social mobility and European civilizational aspirations.

Such aspirations were not just personal, but increasingly national and social. As a Bulgarian national movement emerged in the nineteenth century, the peasant question loomed large, inexorably tied to a newly articulated moral economy of food. Central to the idea of liberation from the Ottomans was the notion that the lifeblood of the peasants (grain) was being siphoned off to feed rich and gluttonous "foreigners" and their local collaborators. Such ideas emerged in conversation with, and were filtered through, the radical politics of the influential Russian populist and socialist left. Bulgarian elites shared with Russians the propensity to rely on a politics of blame for their own poverty and "backwardness." For Bulgaria's intellectual founding fathers, the politics of bread—or sustenance—was tied directly to the notion of freedom from slavery or tyranny. The famous last line of the revolutionary poet Khristo Botev's poem "Borba" (The struggle), "Khliab ili svinets" (bread or lead), is a case in point.[24] Botev's "bread or lead" is a clear reference to the Bulgarian national revolutionary movement's motto, "Freedom or Death," with "bread" replacing (or equivalent to) "freedom." And yet, the issue for most Bulgarians was not so much the *lack* of bread as the problem of a diet *mainly* of bread. Revolutionary politics, then, was tied not just to ensuring bread, but to moving beyond it.

The transition from Ottoman to Bulgarian self-rule in 1878 did little, however, to alleviate the problem. If anything, peasant poverty intensified in the decades that followed. The Russo-Turkish War of 1877–78, in which Russian troops effectively "liberated" Bulgaria from the Ottomans, brought extreme economic disruption and a sharp drop in food production and trade. A wave of Muslim emigration out of Bulgaria freed up large tracts of fertile land, but also wreaked havoc on agricultural production.[25] Many of these lands were hastily sold for nearly nothing or abandoned. As Spas Raikin notes in his memoirs, according to family lore, his great-grandfather had purchased their rather extensive family lands from a fleeing Turk for a *banitsa* (this one filled with cooked pumpkin).[26] The new administration also distributed formerly Muslim-owned lands to Bulgarian peasants or Slavo-Macedonian refugees from Macedonia and Thrace.[27] By 1900, some six million decares of land owned by Muslims had been transferred in one way or another to local Christians.[28] What Bulgarian sources would later describe as an "agrarian revolution" was tantamount to a parcelization of land without an accompanying improvement in agricultural techniques.[29]

Furthermore, after 1878 Bulgaria became increasingly incorporated into the globalizing economy, with grain and other foodstuffs as principal exports.[30] If the Ottomans had provided a source of blame before liberation in 1878, the West—and the local bourgeoisie—emerged as a co-culprit by the late Ottoman period, a trend that only gained momentum after liberation. The slow shift from a subsistence economy to a rural commodity cash economy meant that Bulgaria, especially its rural poor, was vulnerable to a series of global economic crises during the "long depression" of 1873–96. Grain continued to be the dominant export, but the shift to tobacco as a commodity cash crop intensified the potential for peasant hunger and misery. Debt and food prices soared, and for the rural and urban working poor, ever greater percentages of their salaries were devoted to their dietary mainstay. By 1895, prices for flour, bread, and other staples had gone up by 75 percent, and most among the working poor spent about 65 percent of their income for food.[31] New economic vulnerabilities fed the fires of local discontent and increased elite attention to the ever-worsening "peasant" question. For a vocal minority of Bulgarian elites, socialist and agrarian models offered possible answers to the sharpening social inequities.[32]

Bread, War, and Revolution

Bulgaria's spiraling cycle of debt became an important factor in its eventual alliance on the side of the Central Powers in the First World War. The other factor was the desire of Bulgarian elites to acquire neighboring Thrace and Macedonia, which had significant "Bulgarian"-speaking populations but also arable lands under grain cultivation.[33] These desires had brought Bulgaria into the First and Second Balkan Wars of 1912–13. In spite of a net gain of territories in these wars, Bulgaria lost parts of the desired territories of Thrace and Macedonia as well as the fertile breadbasket of southern Dobrudja on the Danubian Plain.[34] The huge addition of war costs to the national debt, along with an influx of more than forty thousand Slavic Christian refugees, brought economic chaos in the aftermath of these wars.[35] Unprepared for another war of even greater devastation, in 1915 Bulgaria nevertheless entered World War I on the side of the Central Powers, with the promise of territorial expansion into Thrace and Macedonia. The war solidified Bulgaria's economic dependency on Austria and Germany, to whom it was compelled to export food and other agricultural goods even when supplies ran low at home.[36] Food scarcity and hunger had also been a problem during the Balkan Wars, but those conflicts were much shorter and localized.[37]

The prolonged nature of World War I meant that food shortages became dire not just in Bulgaria, but all across Europe, as global and regional trade routes and food chains were disrupted.[38] As scholars of food and World War I have noted, the war precipitated unprecedented disruption, plunder, and redistribution of food resources on the part of wartime actors. It also brought a new kind of widely articulated "moral economy of the crowd" from Paris to Petrograd, with anger at authorities and middlemen taking new forms.[39] Tsarist Russia's inability to supply its troops and cities with bread was a key factor in the February 1917 Revolution, just as the continued failure of the Provisional Government to feed the population made calls for "Peace, Land, and Bread" so poignant in the Bolshevik Revolution that October.[40]

For Bulgaria, impending defeat and starvation were a potent mix at the cataclysmic end of the war. Bulgaria had long been a net exporter of food, and growing food shortages were now clearly connected to feeding the distant cities and fronts of the Central Powers.[41] Such shortages exacerbated the constant rise in food prices as a result of wartime speculation that brought visible wealth and abundant food to various local elites and opportunists. Furthermore, Austro-German troops occupying the Balkans received rations that were significantly larger than those given to their Bulgarian counterparts.[42] Since 1915, there had been sporadic protests against food requisitioning and shortages. By the winter of 1916–17, food supplies had become increasingly stretched, sparking growing protest.[43] The Bolshevik victory in Petrograd gave even more credence to the increasingly active Bulgarian Communist Party (BCP), which had opposed the war since its inception.[44] The BCP was by no means the leader of Bulgaria-wide protest, but it actively agitated in Bulgarian cities and among soldiers on the various fronts of the war. By December 1917, around ten thousand Bulgarian workers swarmed the streets of Sofia, calling for peace and revolution. Dimitŭr Blagoev, the leader of the BCP, addressed the angry crowds, calling for an end to the war, but also a halt to all exports to the Central Powers and to local speculation on foodstuffs. He urged the hungry Bulgarian masses to "requisition" food and other items of greatest necessity.[45]

In late January 1918, after food rations were reduced to record lows, the first of what would become a wave of women's revolts broke out in the town of Gabrovo. A socialist deputy to the Bulgarian parliament sympathetically reported that for months there had been issues with the food supply in Gabrovo: "The bread is impossible to eat, full of corn and made from impure and poorly milled flour. . . . It looks like dirt, and even then there is not enough to give out, only 200 grams a person or sometimes none at all, because they are giving out corn on the cob instead of bread or flour."[46] The rest of 1918 was punctuated by women's protests, which grew larger and more organized, and involved attacks on the supply sites of local elites. Urban women took to the streets, joined by rural women who streamed into Bulgarian towns to take part in protests and clamor for bread.[47] In some cases, as in the village of Dolni Dŭbnik and a cluster of villages in the Biala Slatina region, rural women armed with sticks and tools attacked requisition authorities, taking back grain and animals.[48] But the most dramatic protests were in the cities, including Stara Zagora, Plovdiv, Sliven, and Tatar-Pazardzhik.

Appeals to "confiscate" necessary goods were broadcast across Bulgaria through leftist networks, which led to more chaotic scenes as women and children broke windows and doors to take goods from local officials and elites. In the town of Tatar-Pazardzhik, for example, women and children took to the streets for several weeks of protests in July because of a lack of bread. Angry mobs went to the town organs responsible for food distribution, as well as to the homes of local elites. Official reports noted that women from the crowd were heard chanting "Peace and Bread," and more pointedly, "The Germans are robbing Bulgaria, that is why we are destitute, as what they are taking away [food] they will continue to take away."[49] As Blagoev explained in an article in *Rabotnicheski vestnik* (Workers' newspaper), what Lenin had dubbed an "imperialist war" was now a "social war."[50]

The women's food riots and other workers' protests that wracked Bulgarian cities were tied to a larger and more menacing wave of peasant-soldier rebellion that was moving rapidly toward Sofia. These soldiers, under the influence of BCP agitators and propaganda, were keenly aware of their wives' and children's hunger and the protests that had broken out across the home front. By late September 1918, some fifteen thousand peasant soldiers demanding "peace and bread" were caught up in a mutinous wave of desertions, in which a republic was declared in Radomir, a hamlet some forty kilometers from Sofia. As the Radomir Rebellion gained momentum, Tsar Ferdinand was forced to abdicate. But it was the more broadly popular Bulgarian Agrarian National Union (BANU), and not the BCP, that came to the political fore. BANU leader Alexander Stamboliski went on to victory in the first postwar elections, echoing the BCP's promise of bread for the masses.[51] "Yes, dear peasants," the new prime minister purportedly proclaimed, "the world must know that he who provides the bread must be the master and the seigneur!"[52] In the short term, however, Stamboliski was unable to feed the population without help from the Allies, including the United States.

American food aid was critical to not just Bulgarian but also European stabilization in this period.[53] US grain imports had been a key factor in the Allied victory, with food aid to the Allies and the economic embargo of the Central Powers playing a decisive role. In the pivotal year of 1917, the United States entered the war on the Allied side as mass starvation loomed. Given the food crisis on the continent, Woodrow Wilson lifted the blockade on the Central Powers, including Bulgaria, even before the war was over. His aim was to avert starvation and revolution across Europe by advancing a policy of "non-military intervention" via food aid.[54] In 1917 he had created the United States Food Administration (USFA) headed by Herbert Hoover, who mobilized the country's considerable agricultural (and particularly grain) production capabilities. As the war came to an end, the Allied goal of victory over the Central Powers quickly shifted to blocking the spread of Bolshevism across Europe, which was wracked by protest and revolution. As ships laden with the 1918 US wheat harvest set off for Europe, the USFA morphed into the American Relief Administration (ARA), expanding its scope of operations, which led to unprecedented US involvement on the continent. The ARA represented a break from the past policy of allowing private, primarily religious, organizations to be solely responsible for humanitarian intervention abroad.[55]

For the first time, Americans—through private and public agencies—were leveraging food aid not just as a moral imperative, but also as a means of influencing global politics. Private organizations distributed billions of dollars' worth of aid in the period of the war and its aftermath.[56] In 1918 Congress appropriated $100 million in aid, a relatively small amount, but with marked symbolic importance. As the $100 million quickly ran out, the ARA became privately funded, which allowed Hoover even freer rein in distributing food aid across Europe. State assistance still could not be extended to former enemy states like Bulgaria, but with private funds the ARA was freed up to bring food aid to these states. It even supplied grain to Bolshevik Russia, which suffered its first major famine, caused by the disruption of civil war and a devastating drought in 1921–22.[57] Supplying food to a radical communist state was, of course, highly controversial, but as Hoover and other supporters of such efforts rightfully assumed, American food aid, whether officially or privately funded, was a way to broadcast and amplify American economic and geopolitical power.[58] This established the United States as a player in a global game in which food was a critical trump card.

In the late war years and the immediate postwar period, Russia and Eastern Europe emerged as critical battlegrounds in a continental, if not global, politics of bread. As major producers of grain, the peoples and states of the East were the most vulnerable to market fluctuations and disruptions, droughts, famines, poverty, debt, and dependency on more powerful neighbors. They were also the most susceptible to revolution, with intellectuals keenly aware of the deepening poverty not just of the peasantry but of their states as a whole, connected to their semi-peripheral status vis-à-vis Europe and the West.

Bread and Grain between the World Wars

In the aftermath of war, revolutions were averted everywhere except in Russia and its periphery, but the question of bread—if not moving beyond bread—would become part of the national bioimaginary as never before. Like the rest of Europe, Bulgaria entered an era of continual economic crisis and sharpening political and cultural tensions, but also scientific and creative ferment. Both sides of the right-left divide looked for ways to rein in the ravages of unfettered capitalism and ensure that Bulgaria's populations would have their daily bread. The United States and most of the West, after all, quickly retreated into relative isolation, as the Bolsheviks loomed as a specter or a mother ship (depending on one's viewpoint) in the East. Some version of fascism (or at least authoritarianism) slowly crept across Europe from Italy and Germany, subsuming East European politics, not through occupation, but as a political choice. Within or beyond politics, interwar Bulgaria became a rich environment for new modes of thinking and experimentation, political, scientific, and religious, with important implications for bread.

The interwar period brought politics in a new key to Bulgaria. The so-called peasant republic of Prime Minister Stamboliski placed the peasant question front and center in Bulgarian national politics for the first time from 1918 to 1923. Stamboliski directed unprecedented resources to the agricultural sector, seeking not to dramatically mechanize production, but rather to offer peasants credit and encourage the formation of consumer and producer cooperatives. But his regime alienated the left and right alike, and he was assassinated by an amalgam of right-wing forces in 1923, bringing disarray to the Agrarian movement. Beginning in 1923, a series of centrist or center-right coalitions ruled Bulgaria, until Tsar Boris III established a royal dictatorship in 1935. On the left, the Bulgarian communists had orchestrated an ill-fated uprising in 1923, which was handily crushed by the post-Stamboliski regime. The communists (as well as the left more broadly) were pushed out of the political arena and eventually went underground and abroad, or operated through front organizations.

The Bulgarian center and right, like the Agrarians and the left, realized the importance of the politics of bread in interwar Bulgaria. Social tensions and a fear of Communist Party influence in the countryside ensured that Stamboliski's agrarian reform efforts were never wholly abandoned. Ironically, there was actually more redistribution of land—primarily to refugees from Thrace and Macedonia—under Alexander Tsankov's coalition, the "Democratic Accord" of 1923–31, than under Stamboliski's Agrarian regime.[59] There were also efforts to modernize agriculture through the distribution of information and new (albeit basic) technologies—for example, iron plows in place of the old wooden plows—to better feed the population and encourage exports.[60] By the late 1930s, the Bulgarian state also began investing in public bread ovens to boost local supplies of bread. These were not the shared ovens of the past, but rather bakeries that were meant to industrialize the production of bread and free women from the laborious process of making it themselves.[61]

As bread continued to dominate the interwar Bulgarian diet, it became entwined with various political and civilizational questions and narratives. Having enough bread to survive was a politically loaded issue in and of itself, especially given the constant economic crises, which grew in intensity after the start of the global depression in 1929. The findings of a number of interwar studies on the Bulgarian diet, by local scientists and international organizations like the newly spawned League of Nations, confirmed what was already evident to scores of observers: the majority of Bulgarians ate *mainly* bread. More concerning was that the cost of bread was consuming an increasing percentage of the income of poor Bulgarians. Many Bulgarian elites were anxious about the implications of the nutritional (and more specifically protein) gap between rich and poor, but also between Bulgaria and the West. The Bulgarian national diet as emblematic of "backwardness" pervaded interwar discussions of local nutritional needs. Though a solution was far from clear, there was a consistent call for more protein to supplement Bulgarians' bread-based diet. At the same time, what the source of that protein should be, or how to implement such a change in diet, was up for debate.

Bulgarian biochemist and public intellectual Asen Zlatarov was the first to bring food science into academic and public discourse. Notably, he worked not just on the biochemistry of food, though he had trained in this field in Germany, but also on its wider social implications. He had a broader interest, in fact, in larger patterns of calorie consumption among Bulgarians in a national and international comparative context. His most critical findings included a precise accounting of the discrepancies in caloric, and specifically protein, intake between the various classes of Bulgarian society, as well as between Bulgarians and the wealthier nations of Europe.[62] Zlatarov and his assistant Ivan Mitev conducted extensive research on the diet of Bulgarians of various social classes. Drawing on an array of earlier studies of food in the household budget, they documented the poverty of the lower classes as reflected in their paltry caloric intake and household budgets and their largely bread-based diets.[63] They also concluded that Bulgarians as a whole ate considerably fewer calories and took in lower amounts of protein and vitamins per capita than citizen of countries in the industrialized West, a gap that seemed to be widening.[64] For Zlatarov, the Bulgarian bread diet was a clear indicator of the masses' widespread poverty, and hence the nation's "backwardness" in relation to the West.

And yet Zlatarov did not merely appropriate all of the (ever-changing) assumptions of Western science. Nor did he simply assume that the Western diet was somehow superior. He recognized that more protein was needed in the Bulgarian diet, but also believed that many Western studies had overestimated the need for protein. With that in mind, he refuted the notion that the average adult body required 120 grams of protein per day, concluding that 60 grams was sufficient.[65] In addition, Zlatarov did not categorically reject all aspects of the "traditional" Bulgarian bread- or grain-based diet. In fact, he noted with alarm that the problem was not bread per se, but rather that "European" forms of white flour and bread were becoming more common in interwar Bulgaria. Metal roller mills had replaced stone gristmills in the United States and much of Western Europe in the 1880s, producing new flours in which the wheat germ and bran were sifted out. These technologies and tastes had spread to Bulgarian cities by this period, where white bread was increasingly connected to wealth and status but also consumed by the urban masses. Zlatarov lamented the fact that the new forms of white flour and bread were stripped of so many nutrients, including protein and newly discovered vitamin complexes. In addition to advocating that the nation hold on or return to whole grain bread, Zlatarov believed the Bulgarian diet would benefit from an influx of a wide variety of vegetables and fruits and plant-based proteins like legumes, along with increased dairy and eggs. Far from proposing a simple imitation of the Western meat-based diet, Zlatarov, who was a committed vegetarian, looked in part to the global vegetarian movement (explored in more depth in the next chapter) for a more sustainable model that made sense for Bulgarians. His own research confirmed that a sustainable plant-based dietary system was healthier, but also made the most fiscal sense for Bulgaria, given its relative poverty.[66]

For Zlatarov, who had socialist sympathies, the "problem" of Bulgarian food called for a science-based nutritional as well as socioeconomic solution. He looked to the Soviet Union, where, from his perspective, science was well supported and mobilized for the good of society. Traveling to the USSR in 1936 for a scientific congress of physiologists, he met and conversed with the famous Russian scientists Nikolai Vavilov and Ivan Pavlov—important figures in Russian, if not global, food science.[67] In his travelogue about this journey, Zlatarov described strolling through Vavilov's Institute for Agriculture and then sharing a meal with the renowned plant scientist. At the dinner, Vavilov described his Soviet-supported world travels, from Central Asia to South America, in search of seeds, roots, and whole plants; he amassed a collection of hundreds of thousands of live specimens, with the lofty goal of averting famine in the Soviet Union, but also feeding the world.[68] The Soviet Union offered a glimpse of a future in which science would be elevated to new heights, underpinned by a redistribution of wealth in the service of feeding the masses. Zlatarov was fascinated by the ways in which the Soviet state was wielding power through an explicit biopolitics—namely, its coordinated effort to organize and optimize collective human potential and performance.

One can only presume that he was unaware of the darker side of the story of Stalinist biopolitics: how Stalin used hunger and starvation as a political tool during the forced famine in Ukraine in 1932–33. Nor did Zlatarov witness the infamous wrong turn in Soviet plant biology when, shortly after his visit to the USSR in 1936, Stalin chose to favor the plant genetics of Trofim Lysenko, or what came to be known as Lysenkoism. With Stalin's support, Soviet scientists were required to adhere to Lysenko's notion of the heritability of "acquired" characteristics, in contrast to the more widely accepted notion of biological inheritance associated with the nineteenth-century Austrian scientist Gregor Mendel. Lysenkoism crippled agriculture in the USSR and later the Bloc from roughly 1936 to 1953, as thousands of Soviet proponents of Mendelian genetics perished in the Gulag for their views.[69] Vavilov was among them. Arrested in 1940, he died of starvation in prison in 1943, after a lifetime devoted to feeding the world.[70] Zlatarov did not live to see the Soviet wrong turn in plant science or the arrest of Vavilov. His own affinity for the Soviet experiment brought with it surveillance by the right-wing Bulgarian state, driving him into self-imposed exile in Vienna, where he died in 1936. He left behind his extensive writings, as well as the students and other thinkers he influenced—in short, a legacy of homegrown food science and biopolitical thinking framed by Marxist thought. He planted a seed in the Bulgarian bioimaginary, whereby science might replace religion.

For other interwar thinkers and movements, however, an embrace of food science did not necessarily eclipse the spiritual. For example, the Bulgarian Tolstoyan movement—which originated at the turn of the century—gained in strength after World War I. The members of this pacifist and avowedly vegetarian movement were focused first and foremost on the original religious and ethical ponderings of the famous Russian author Lev Tolstoy, who had died in 1910. Bulgarian Tolstoyans, many of whom lived in agricultural communes, had also long been involved in publishing, printing the works of Tolstoy and other works on nonviolence and vegetarianism. By the interwar period, their publications included periodicals like *Vegetarianski pregled* (Vegetarian review), whose articles introduced new advances in food science to bolster their advocacy of a plant-based diet. If religious thought was still critical to their ponderings on avoiding meat, in their writings on bread they, like Zlatarov, were focused on the science of whole grain. *Vegetarianski pregled* was filled with articles, translated from English, German, and French, that were critical of white bread as a product of the "ill-informed but highly civilized human race."[71] Tolstoyans connected white bread to modern technologies and ways of life, the ills of civilization that they sought to escape on their communal farms. And yet they also employed contemporary science to prove the superiority of whole grain; science seemed to confirm (and not repudiate) the meaning and presence of God in bread. If Zlatarov was more concerned that bread be whole than holy, for the Tolstoyans whole grain was indicative of God's bounty that modern man had marred.

The latter also held true for another prominent interwar Bulgarian movement, the Bialo Bratstvo (White Brotherhood), or "Dŭnovtsi." The latter name came from their founder, Petŭr Dŭnov (aka Beinso Dŭno), who attended American Protestant missionary schools in Bulgaria and then went on to study theology and medicine in Boston from 1888 to 1895. After he returned from his studies, a spiritual community formed around him, which gained in size and scope amid post–World War I soul-searching. An exponent of "neo-theosophy"—a kind of gnostic pan-spirituality, heavily influenced by yoga and focused on open-ended spiritual connection with "cosmic energy"—Dŭnov came to be known as "Uchitel"—the Teacher. His ideas, which he propagated through his famous "conversations," clearly referenced those laid out by the theosophical movement founded by a Russian expatriate in the United States, Helena Blavatsky, in 1875. Blavatsky was enamored with the Tibetan Buddhist myth of the hidden utopian world of Shambala and the "Great White Brotherhood" of enlightened beings who inhabited it. The White Brotherhood was named in clear reference to these beings, and Buddhist thought was foundational. Still, contemporaries characterized Dŭnov as an original and profound thinker whose wisdom was purportedly sought out by luminaries of the period such as Albert Einstein and Tolstoy. Although it is hard to separate mythology from fact, Bulgarian sources seem to suggest that when Tolstoy left his home in his final days, he was on his way to Bulgaria to meet Dŭnov.[72] Interestingly, Bulgarian Tolstoyans made similar claims.[73] It is not inconceivable that both are true. Alas, Tolstoy died along the way.

The White Brotherhood's alternative vision (or critique) of the modern world, including vegetarianism and natural healing through food, attracted a large number of followers and sympathizers, including well-placed cultural and administrative elites. Among the latter were Evdokia, the sister of Tsar Boris III, and two of the tsar's closest advisers, Liubomir Lulchev and Ivan Bagrianov—as well, presumably, as the tsar himself.[74]

FIGURE 1.2.   Asen Zlatarov in Paris, 1926. TsDA, F-865K, O-1, ae-495 via Wikimedia Commons.

By 1921, Dŭnov and his followers had established the Izgrev (Sunrise) commune on the outskirts of Sofia. Here a community of followers, dressed all in telltale white, lived in a communal village of small houses surrounded by gardens where they grew food for the collective.[75] Plant-based foods played a special role for the brotherhood, and wheat had a particularly holy status. In the words of Dŭnov, "each wheat grain bears consciousness and Divine thought."[76] The science of food was also present in his influential teachings, but as a kind of abstraction. For example, in Dŭnov's conversations one can find such pronouncements as "I can prove to the scientists that Christ is hidden in wheat grain," and "Wheat grains contain potential energy which transfers to kinetic when entering the human organism."[77] But Dŭnov's concerns were less about vitamins and minerals and more about how "the essence of wheat is hidden in this element that the chemist cannot find."[78] Instead he embraced it as "the best food," declaring that "man can live on bread and water only."[79] In line with Zlatarov and the Tolstoyans, Dŭnov also decried modern milling practices, noting that "bread as done nowadays has no great nutritious value because prana has gone away."[80] For the Dŭnovtsi, only whole bread could be holy.

The cause of whole grain bread was clearly ingrained in a variety of thinkers and movements whose concern was the preservation, if not renewal, of the collective national body. Such projects crossed the secular-religious as well as the left-right divide. The foreword to a right-wing 1940 pamphlet titled *White or Black Bread*, for example, argued, "For the Bulgarian nation, the bread question is of life-and-death significance, because bread is our main food. . . . It is bread that gives strength and stamina to the peasant, worker, and soldier. It has fortified us for centuries, and will scaffold our health, and life in the future. . . . Everything else that had fed the national masses is insignificant and merely for flavor."[81] As the pamphlet's author, Petŭr Georgiev, elaborated, whole grain bread, which contained "all the necessities of life," was being replaced by white bread, which was sullied by the modern "fashion" of milling. As a result of such practices, it was "devoid of bran, protein, and other elements."[82] This, he argued, was the root cause of malnutrition and chronic hunger, particularly for the poor, as only the rich could afford to supplement their bread diets with other foods.[83] For Georgiev, the implications were clear: "The fashion of white bread is a losing prospect for the health and strength of the Bulgarian nation." He called upon doctors, teachers, and poets—"those who are closest to the nation"—to come out against white bread, which had turned Bulgaria into a "weak nation" and kept it from taking its place in the "family of nations."[84] Significantly, Georgiev looked to "cultivated nations" such as Nazi Germany, where an organized movement had embraced a turn to whole grain breads over white.[85] He claimed, in fact, that the Germans had "outlawed" white bread.[86] In truth, however, the Nazis had only launched a campaign against industrialized white bread, for reasons of its high cost (white bread was still a luxury) and lack of nutritional benefits.[87] Georgiev's concerns for the Bulgarian nation, while not fascist, were nevertheless inspired by a fascist fascination with the body, as well as concerns for the survival and success of the national body politic.

The urgency of ponderings about food and the future of the nation only intensified as the shadow of war again moved across the continent. A range of food narratives, vegetarian and otherwise, called for a better-fortified nation via a diet of whole grains, but also ample vegetables and fruits (and for some, meat).[88] In the words of Zakhari Ganov, editor of the periodical *Khrana i zhivot* (Food and life), "a well-fed man can be a productive worker, a strong soldier, a good father, and a loving spouse." But as he also noted, food and markets were the main reason for the current war, in which the primary concern of each state was feeding its population.[89] The memory of the mass hunger brought about by World War I was still fresh, but so too were the unhealed wounds and unfulfilled territorial ambitions.

Bulgaria entered World War II on the Axis side in 1941, in large part because of its dependency on Germany as a market for its interwar exports. Grain was only one of those exports, but it was a critical factor in Bulgaria's close ties to Germany. In fact, Nazi Germany had been buying grain from Bulgaria at above-market prices since the 1930s, as part of its effort to stockpile food supplies at home for the impending war.[90] Bulgaria's alliance with Germany also brought the promise of expansion into the agriculturally rich southern Dobrudja and parts of Macedonia and Thrace. For Bulgaria there was a clear benefit to the alliance, and yet in some respects it became an agricultural colony of wartime Germany, exporting food at the expense of local consumption.[91] The Germans had the upper hand in setting the terms of trade and largely controlled the flow of goods within "New Europe"—their growing patchwork of allied, quisling, and occupied territories. Germany's domestic food production and exchange, as well as the policy it implemented for the continent as a whole, was highly calculated. In essence it was a "total food policy," based on removing Germany from the wider global food system and creating relative self-sufficiency, integrally tied to import treaties with Southeastern Europe—namely, Romania, Bulgaria, and Yugoslavia.[92]

German policy was also grounded in its own intensive biopolitics, in which food was a weapon and Europe was a "larder" for "organized hunger." This divided Europe into the well-fed, the underfed, the hungry, and the starved. Racial hierarchy was key in this regard, as the "lower races" of Eastern Europe (primarily Slavs and Jews)—occupied or allied—were starved out at the expense of Germans and other higher "races" in the West, who were afforded higher rations.[93] "Bleeding the Balkans" was undoubtedly part of the Nazi strategy, as Bulgaria and its neighbors were required to make enormous deliveries of grain and meat along with agricultural workers, even as their own bread rations continually decreased. In addition, Germany urged Bulgaria and its allies to conserve and to use coarser fodder grains for human consumption, even as the Germans themselves had the highest rations of wheat and meat in Europe.[94] Exports of food to Germany contributed to shortages, hunger, rationing, and speculation.[95] At the same time, because of intensive new methods of agriculture, as well as late interwar stockpiling, Germany was less of a drain than in the previous war. The worst Bulgarian grain shortages were caused by a disastrous drought in 1942, after which Germany actually supplied wheat to its southern ally.[96] Bulgarians, like other "allied" Slavs—particularly the Czechs and Croats—had rations roughly equal to those of the Italians and French.[97] That is to say, Bulgaria occupied a relatively comfortable place within Germany's orchestrated "geography of hunger."

In addition, the wartime state was able to avert mass hunger through a more direct and effective state coordination of the food system. World War I and the economic crises of the interwar years had driven a growing political sense (in Bulgaria and beyond) that food supplies should be managed by the state. During World War II, the Bulgarian state was much more hands-on in coordinating food production through Khranoiznos (literally "grain export"), the state agency that controlled the grain trade from 1931 to 1948.[98] With tighter management of the economy, the dire circumstances of the final years of World War I were avoided. Still the war ended in loss, another national catastrophe for Bulgaria in several respects.

When the advancing Red Army occupied Bulgaria in September 1944, a nominal "people's revolution" took place. In contrast to 1918, however, this was not a spontaneous uprising of the hungry masses, but rather a "revolution administered" by the newly empowered BCP, who would bring about a revolution of another kind in the years that followed, in and beyond bread.[99]

Cold War Bread

The first order of business for the postwar regime—along with purging and punishing "fascists," very broadly defined—was to provide bread to the population. The country was not starving, as it had been in 1918, but there were severe food shortages, and the economy had been crippled by the war effort. From 1944 to 1947, the BCP played a dominant role within a fragile and largely nominal multiparty system. After 1947, only the Agrarian Party would remain in place alongside the BCP, albeit in a subservient and pro-communist form. As part of the political transition to a functionally one-party system, in 1944 key elites from the wartime period were tried in the infamous "people's courts." This included high-level government officials and advisers in virtually all sectors of the economy, including the four officials who had coordinated the state monopoly on grain exports. All four were convicted as profiteers who had furthered the German and Italian war machines at the expense of the Bulgarian national economy.[100] As the BCP leadership argued, the newly imposed national control of resources meant that the nation would no longer be held captive by foreign interests. Instead, under the new system, local but also "rational" management would be possible under Soviet tutelage, saving Bulgaria from past "waste and abuse."[101]

Food was a critical component of the new regime's drive to establish its legitimacy, consolidate power, and reorder society. This was a far-reaching effort, in which agriculture was only one part. Nevertheless, the grain economy and bread consumption were important elements of the communist project from beginning to end. The Fatherland Front government took over the wartime state monopoly on grain trade and exports as early as 1944, a full two to three years before it imposed a monopoly on the trade in other goods. Also in 1944, the regime began the process of collectivizing agriculture and promoting public bakeries in place of home baking.[102] The goal was to ramp up production but also to stabilize and heavily subsidize the price of bread along with other food prices. Providing centrally produced, cheap bread was seen as one of the best means available for the new regime to consolidate power.[103] Public production of bread would also "liberate" women from the time-consuming task of making it themselves.[104] To some extent, small-scale bakeries had already accomplished this in Bulgarian cities, but not in rural areas, where the bulk of the population still resided. There all women still made their own bread, which in communists' calculations was an irrational use of time and labor. The estimated five to six hours a week could be better used in agriculture, industry, or other sectors that contributed to the larger project of "building socialism."[105]

The transition to socialism took place in the larger context of the rapidly chilling Cold War, where a competition for bodies and souls cast its shadow across a devastated Europe. In this contest, grain supply was a central concern, and the United States, in contrast to its rapid retreat after World War I, made it clear via the Marshall Plan that it was in Europe for the long term. With the Axis powers defeated, the United States and its allies saw Soviet-backed communism as the primary threat to the continent as the Soviet Red Army occupied Eastern Europe. Tensions among the Allies over postwar territorial and political configurations had been brewing since 1943, and now the war for territory was being fought, as in 1918, via the politics of bread. The United States had supplied food to its allies, including the Soviets, since 1941 via the Lend-Lease program. This continued through the United Nations Relief and Rehabilitation Administration (UNRRA), which was formed in late 1943, predating the United Nations. The largely US-funded UNRRA provided food and other supplies to the continued war effort as the Axis forces were rapidly pushed back over the course of 1944. Once the war was over, massive relief continued—three times what the United States had supplied in the post–World War I period.[106] Just as after World War I, US aid was connected to the justified fear that hunger, which was acute in large swaths of Europe and Asia, might fuel communism's spread. Aligned with President Harry Truman's emergent containment policy, the UNRRA targeted aid to Greece, Italy, and Yugoslavia, where local communist movements were particularly strong.[107] As the Cold War set in, however, the UNRRA became a liability, as it was nominally an international body still tied to cooperation with the Soviet Union. The reality of communist takeovers and the threat of communism's spread pushed the American leadership to launch a separate and expanded food aid program as a critical tool in shoring up Cold War battle lines in Europe.

By 1947 the United States had rolled out the European Recovery Program (also known as the Marshall Plan), in which food credits and the rebuilding of agricultural capacity in Europe were central. As George C. Marshall noted in his famous Harvard speech in 1947, "Our policy is directed not against any country or doctrine, but against hunger, poverty, desperation, and chaos."[108] The 1945 birth of the United Nations as a multilateral international organization brought with it the establishment of the UN Food and Agriculture Organization, which was set up to monitor and supply information on global food production, consumption, and exchange.[109] Efforts to eradicate global hunger by gathering and dispersing information on global food production and nutritional needs were not always explicitly political. And yet, many in the West feared that "desperation and chaos" would lead states inside and outside the Soviet occupation zone to turn to communism. Capitalism, after all, had not proven to be a system that could provide stability and prosperity for all. In recent memory, it had brought global depression, vulnerability, dependence, political chaos, war, and hunger.

American Marshall Plan aid was offered to all of Europe, including the Soviet-occupied states of Eastern Europe. Under the USSR's direction, these countries—which were in various stages of a communist takeover—turned down the proffered aid on the grounds that it was tied to "American economic imperialism."[110] Stalin was well aware that the newly forming Eastern Bloc now contained the continent's primary breadbaskets, but that many of these states were in ruins and in desperate need of aid. The USSR's efforts to resuscitate its own agricultural production in 1946–47 had paid off, and 1947's harvest was even better. This enabled the Soviets to implement what came to be called the "Molotov Plan," after the then minister of foreign affairs, Vyacheslav Molotov. A cornerstone of the plan was the use of grain to strengthen the Soviet position in Eastern Europe. In many cases, most notably in portions of industrialized Central Europe, Soviet grain imports coincided with the mass "export" or expropriation of machinery and equipment. In Bulgaria, however, grain shipments were a welcome input into a "backward" economy, where Soviet expropriation was less possible or less common. Some 350,000 tons of Soviet wheat was shipped to Bulgaria in 1946 alone, with a promise that the "gift" would not have to be repaid.[111] Soviet grain bolstered the postwar communist regime, which was able to rather quickly and efficiently assume power, using both the stick of organized redistribution of wealth and the carrot (or loaf) of social advancement and economic recovery. The Molotov Plan was accompanied by the formation of the Communist Information Bureau (Cominform) on October 5, 1947. This organization was central to the emergence of the Eastern Bloc; its establishment was a declaration of an economic and political counterweight to the Marshall Plan. By 1949 the Council for Mutual Economic Assistance (Comecon) would take over the function of Bloc trade integration, coordinating what became a kind of separate global trade system that was hermetically sealed, at least from 1947 to 1956.

At the same time, within these states, the imposition of a new political and economic system had far-reaching consequences for food production. The collectivization of agriculture proceeded apace, albeit in fits, starts, and in some cases stops. Collectivization was a central—if not *the* central— component of the Bulgarian socialist project, affecting 75 percent of the population.[112] As the prominent historian of Bulgaria Richard Crampton noted, this process amounted to "the greatest social transformation forced upon the country since the Ottoman conquest." Prior to collectivization, four-fifths of the population, some 1.1 million people, had earned their livelihoods from small farms, on average under four hectares.[113] Wheat and other grains were still sown with a wooden or iron plow pulled by oxen or horses, then harvested by hand with a scythe. There were the beginnings of change, to be sure, but even the iron plow was a recent import. New ideas and methods were slowly percolating, but change was minimal. If anything, farms had become smaller and smaller, as population growth and an influx of refugees had exacerbated land hunger since 1878. At the same time, there was a strong foundation for the idea of the "cooperative" farm, which had put down roots in Bulgarian soil at the turn of the century. The Bulgarian cooperative tradition was tied to Agrarian, but also socialist, Tolstoyan, and other pre–World War II movements.[114] To tap into the affinity for these traditions, the BCP referred to collective farms as "cooperatives," or more precisely *trudovo kooperativno zemedelsko stopanstvo* (labor cooperative farm).

Collectivization had a mixed record of successes and failures, costs and benefits. The new cooperatives, the BCP promised, would be voluntary like the old ones: their members would share the profits, and could leave at any time. In practice, from 1945 to 1947 it was primarily the poor or landless peasants who joined voluntarily.[115] The new system of agricultural organization was most advantageous for those with nothing to lose, as the collectives had better access to seed and supplies. By 1949, only 11.3 percent of all arable land had been collectivized, and resistance seemed to be mounting.[116] For that reason, a more coercive process of forced collectivization began in 1949 and continued until 1951. As in the Soviet case, the primary target was so-called kulaks, broadly defined as any farmer who used paid labor, which amounted to about 25 percent of the peasantry.[117] Another group affected by this wave of collectivization was Bulgaria's Turkish minority. On the heels of the 1950 "Six-Year Plan," the state created vast collective farms in the grain-rich Dobrudja, and 140,000 Turks were subject to a large-scale expulsion.[118] As a result of intensified collectivization campaigns, 60.5 percent of all the arable land in Bulgaria (and 65 percent of the arable land in Dobrudja) had been collectivized by June 1952.[119] The process was painful, to be sure, and peasant resistance was so great that the Communist Party was compelled to ease up on this forced "shock" campaign, and even acknowledge its own mistakes in 1951. After Stalin's death in 1953, party members were even more harshly subjected to "self-criticism"—a common practice under communism. Stringent measures that forced peasants off their land in effect had already shattered rural resistance, without immediate economic reward. Small parcels of land (and along with it, rural labor) had been consolidated, but collectivization had merely created hundreds of larger farms that were still worked by "manpower and slow-moving buffalos [or oxen]."[120] The disruption, as well as the BCP's focus on heavy industry, led to continued rationing, bread lines, and the need for imports from the USSR in the short term. Grain supplies remained inconsistent, subject to drought, distribution problems, and pilfering—or supposed disruption by "kulaks."[121] To be fair, food rationing continued in much of Europe in the late 1940s and even into the early 1950s, in spite of the Marshall Plan.[122] And in Eastern Europe, like the West, the postwar recovery slowly took hold.

Overambitious state production quotas were not always met in these years, but organizational changes—however painful—eventually did bear fruit. In Bulgaria the collectivization process created the conditions for more consolidated and modern farms, where agricultural technologies could be applied, labor could be mechanized, and production could be exponentially increased. This was particularly true in terms of wheat and other grain production in the fertile valleys of the Dobrudja and Thrace. The success and failure of the system would continue to be determined by its ability to provide bread, but increasingly success was measured not by bread alone.

Communist Bread and Butter

The postwar recovery was on more or less solid footing until Stalin died in 1953, which opened the door for dramatic changes within the Bloc. This included a shift from heavy to light industry and services, and a focus on consumer goods, including a cornucopia of food products and service venues. It also meant a slow opening to trade and technological exchange with countries outside the socialist camp. Within agriculture, this shift meant greater attention to the industrialization of all phases of food production. As early as 1953, then premier Georgi Malenkov introduced a set of reforms to ramp up agricultural production through increasing price incentives to collective farms, but also the encouragement of food production on private plots.

In the same year, Nikita Khrushchev, who was soon to solidify his position as Soviet leader, rolled out his ambitious "Virgin and Idle Lands" program, which spurred the cultivation of vast new territories in Kazakhstan, Siberia, Ukraine, and southern Russia.[123] It was clear to Khrushchev and his generation that the superiority of the socialist system had to be proven not just in steel, weaponry, and space, but also from farm to table. Under the watchful eyes of the West, Khrushchev denied that such changes were being implemented in reaction to a "bread crisis" or generalized food shortages. In fact, he explained, the immediate postwar bread problem had been solved. Instead, the reforms were a response to increased demand resulting from the rather explosive growth of urban populations needing to be fed.[124] Contrary to his statements, there were continued shortages, at times bordering on crises, caused by problems in the production and distribution chain. But Khrushchev was also correct in assuming that urban populations were ballooning; an army of socialist consumers was emerging across the region, with ever higher expectations for consumer goods.[125] With new agricultural lands, methods, and technologies, a steady supply of bread and other wheat-based products could be ensured for the Soviet Union and the Bloc, as well as other global clients. At least partial success in the Virgin Lands program from 1954 to 1958 allowed for Soviet exports of food to the Bloc to curb real and potential resistance—such as the protests (and even revolution in Hungary) that followed the 1956 denunciation of Stalin's crimes.[126] In the post-Stalinist era, socialist states needed to feed burgeoning socialist cities and provide a palliative for protest, as well as evidence of the promised socialist "good life"—which was a way station on the road to the utopian communist future. After all, what was the use of launching Sputnik (in 1957) if the shelves remained empty?

Khrushchev was well aware that the United States was anticipating a massive agricultural surplus because of the rapid expansion of new agricultural methods, from chemical fertilizers and pesticides to hybrid seeds and monocropping. He also knew that US president Dwight D. Eisenhower had introduced the 1954 Food for Peace Act, which diverted US agricultural surpluses abroad in order to stimulate trade, feed the world, and further the objective of "containing" communism globally.[127] In response, Khrushchev was determined to *surpass* the West in agricultural production by 1960, and even to reach "ripe communism" by 1980. His new foreign policy of "peaceful coexistence" with the West was coupled with ambitious plans to outperform and overtake the West while building a better and fairer system. Food was an essential cog in the wheel of this system.

Khrushchev had no qualms about using Western science and technology in food production to achieve Soviet and Bloc objectives. As he noted in his memoirs, "We need to learn from the capitalists, as Lenin urged," but use their technologies to enrich "the people" as opposed to just the bourgeoisie.[128] To this end, in the 1950s Khrushchev approved one of the most important shifts in Soviet agronomy, with particular implications for grain. The official rejection of Lysenkoism meant that Mendelian genetics (and the ideas of Nikolai Vavilov) could again be used in plant science, which would be a boon for the production of hybrid seeds. In combination with his Virgin Lands project and the expansion of "agro-cities" or "agro-industrial complexes," this contributed—for better or worse—to the growing colossus that was Soviet and Eastern Bloc agriculture. Ever-expanding production quotas required ever more mechanization to feed Bloc populations, who were hungry for the promised "good life." At the same time, socialist states continually marked the parameters of this good life, namely through expectations of "rational consumption."

Since the immediate postwar period, such "rationality" had always been framed by the guiding principles of nutritional science—which now became a prerogative of the state. Interwar "progressive" food science—namely, the ideas of Asen Zlatarov—offered a potential model for postwar nutritional standards, but with important revisions. Recall, for example, Zlatarov's view that whole grain bread was nutritionally superior to white. These ideas were echoed in the work of others in the interwar period, including Zlatarov's student and associate Ivan Tomov, who was also a biochemist. In 1945–46 the Ministry of Health commissioned Tomov to write a pamphlet titled *Something We Should Know about Our Eating Habits*.[129] In it Tomov clearly laid out the importance of "rational eating," with nutritional science as a guide for the individual and collective national diet.[130] He called on readers to make use of new kinds of scientific knowledge and frame their eating choices around a proper biochemical balance of nutrients, with a plant-based diet consisting of plenty of black bread, fruits, and vegetables, and no alcohol.[131] Tomov's proposed model of dietary restraint made sense from a socioeconomic and nutritional standpoint.

But such requirements were in competition with the need to generate political consent for the new system among the Bulgarian population. The question was, would the postwar era be one of continued dietary austerity, as Tomov proposed? Or would it be a period in which a bounty of foods would be available to all—a utopia of plenty? For the socialist state, it was a constant struggle to find a balance or synthesis of the two. But in this period, Tomov's pamphlet proved to be too austere in its message for the emergent postwar regime, which severely limited its distribution and promotion. As Tomov later noted with dismay in his unpublished autobiography, the Ministry of Education informed him that the pamphlet did not "harmonize with the Fatherland Front's ideology of food."[132] His focus on whole grain bread was one of the disharmonies. If whole grain bread was superior from a nutritional perspective, in other respects it created a problem for the communist state. For one thing, refined white flour was far more stable and easier to use in industrial factory settings. White bread made sense from the standpoint of fulfilling state plans for the mass production of bread and bread products. But white bread also satisfied the regime's need to show "progress" in terms of the "standard of living," which was at the heart of the socialist promise. In the past, leavened white bread had been a special holiday food for most—too time-consuming and costly for regular consumption. It was associated with the urban, the civilized, and perhaps also the (real and imagined) West. In the communist period, white bread had important symbolic value, and so it became the order of the day in spite of its nutritional deficiencies.

The mechanization of bread production from farm to table was seen as nothing less than a panacea for the "irrational" ways of the past. Notably, early Bulgarian communist-era cookbooks did not include recipes for bread. Perhaps such basic instructions were not needed for those with bread-making skills, but the message was clear: bread was to be bought, not made at home, which was deemed a waste of women's time. Prior to 1944, the bread "industry" in Bulgaria had been limited to small-scale "craft" bakeries with wood-burning ovens in interwar cities. As the BCP took control of the national economy, it moved to immediately build rudimentary bread factories in Sofia, Plovdiv, Pernik, and Varna. These were then the primary source of industrialized bread until 1960, when some forty new bread factories made it into the ambitious new economic plan, to be built by 1980. These factories, along with a bevy of new specialists, were incorporated into the production process to use science and technology to bring white bread to the Bulgarian nation.[133] Bread making was now seen as a process that required expertise in agronomy, microbiology, biochemistry, and other sciences. As part of the mechanization of production, the pursuit of "quality" meant a massive shift to the use of refined white wheat flour for bread, as opposed to the then common mix of "low-quality" rye, corn, and wheat flours.[134] Of course, "mechanization" only went so far. As late as 1970, 93 percent of Bulgaria's bread ovens were still wood-fired, while only 30 percent of the dough for factory-produced bread was machine-mixed.[135] According to official reports, equipment was old and worn-out and was unable to meet demand from Bulgarians, let alone visitors in resort areas like Varna.[136]

Demand was the new order of the day, as socialist consumer expectations rose as high as a freshly baked loaf. Bread and pastry consumption continually went up, as did consumption of a variety of other bread products, now increasingly laden with refined white flour, dairy products, and sugar. Decadent pastries like *banitsa* were more widely baked at home and prominently featured in cookbooks. They were made available in kiosks as street food, as well as in shops. What had once been regarded as "holiday" food was now everyday fare, another kind of socialist triumph. White bread and buttery, sugary pastries were a harbinger of both systemic success and a turn toward a "modern" diet, for better or for worse. Such a transition was another aspect of the socialist quest for progress, of "catching up to the West." The goal remained for all Bulgarians to eat less bread as a percentage of their dietary calories. But "progress" also demanded a shift to white bread and white flour pastries in all their many forms, even given the clear nutritional cost.

The story of bread in the United States offers a stark example of the extremes—and the costs—that the industrialization of bread could take in the Cold War period. In the United States the ability to convert flour into fluffy white loaves on a massive scale was envisioned as a kind of triumph of the modern age beginning in the late nineteenth century. Cheap mass-produced white bread was a central product and driver of the industrialization of agriculture and food processing in the United States, which reached its apex in the twentieth century in the shadow of World War II and the Cold War.[137] The Cold War witnessed the triumph of the US factory-made, pre-sliced, shelf-stable, vitamin-fortified, and plastic-wrapped loaf of Wonder (and other bread brands). Its central place in the American diet was connected to the belief in the superiority of industrial food, as well as the capitalist system. But, as US food historian Aaron Bobrow-Strain poignantly asks, "Is this stuff even food?"[138] Indeed, this form of bread was practically devoid of naturally occurring nutrients, and it was actually detrimental to health in a period when obesity and chronic diseases like diabetes were skyrocketing in the United States. Certainly there was a connection. As Bobrow-Strain argues, in the United States white bread came to symbolize both "the apex of modern progress and the specter of physical decay."[139] Nothing along the lines of American Wonder Bread appeared in Bulgaria in the socialist period. Nevertheless, there were some parallels. The shift to mass-produced white bread could certainly win over the masses, but at what cost?

Against the Grain?

Fulfilling the population's needs and desires for bread and bread products was an important element of the communist promise. But in certain respects, bread became an increasing burden, even a liability, in terms of its place in Bulgaria's nutritional calculus. In 1959 the Institut na khranene (Institute of Nutrition) was created, with the goal of improving the "rationalization" of food production and consumption.[140] The institute and its satellites employed a fleet of scientists, engineers, agronomists, and cooks, who studied everything from the biochemistry of foods to the mechanization of production and packaging techniques.[141] A range of food scientists tracked data that showed the Bulgarian diet improving each year, becoming more diversified in terms of meat, milk, and fruits and vegetables. And yet there remained serious concerns that Bulgarians were getting too high a percentage of their calories from bread and flour products out of "habit and tradition."[142]

As a follow-up to the research of Zlatarov and other interwar scholars, the Institut na khranene launched a range of studies to precisely track the caloric and nutritional aspects of Bulgaria's "transition to socialism."[143] The general consensus was that by 1962, almost all Bulgarians had achieved a scientifically determined level of necessary caloric intake, but they still ate too much bread.[144] Bread and flour consumption had decreased as a *percentage* of the diet since the last study in 1939, but Bulgarians still consumed between 251 and 275 kilograms a year in 1959, well over the "rational norm" of 180 to 190 kilograms. It was also not far below the 1938–39 level of 286.9 kilograms.[145]

The bigger issue, however, was how Bulgaria compared in this respect to other socialist countries like Yugoslavia (187 kilograms) and Poland (142 kilograms), or neighboring countries like Greece (167 kilograms). The difference was even starker in relation to the "West," for example France (107 kilograms) and Austria (113 kilograms).[146] The implications of the numbers were clear: if Bulgaria wanted to "catch up" to its Bloc rivals and the West, its citizens needed to eat less bread.[147]

Producing enough wheat to satisfy local demand was part of the problem. On the one hand, by the 1960s Bulgaria was leading the Bloc in economic growth indicators, with 12.2 percent growth in agriculture (as compared to the USSR at 10 percent and Hungary and Poland at 5–6 percent).[148] By 1967 Bulgaria had become a net exporter of grain and was one of the few self-sufficient or "self-feeding" Bloc states that could actually shore up Bloc grain supplies.[149] On the other hand, agricultural industrialization was uneven, ever plagued by problems in storage and transport and a lack of fertilizer and expertise. There were intermittent problems in planning and distribution, along with bad harvests. This, along with climatic conditions (especially drought) and a series of bad harvests, made Bulgaria and the Bloc as a whole, as an integrated and still largely separate food system, vulnerable. Beginning in 1963, bad harvests in the Soviet Union reverberated throughout the Bloc and its food system. The Soviet Virgin Lands program, while not without successes, had resulted in catastrophic dustbowl conditions in Kazakhstan, with unforeseen drought and massive erosion. While corn was plentiful, bread lines appeared again in the USSR, which created uncertainty about grain imports to the Bloc. Wheat shortages in Bulgaria, as a result of insects, floods, and hailstorms, meant that there were bread lines there, too, and bread was being made from a mix of cornmeal and other grains, like barley. Local communist parties and left sympathizers refuted this reality. For example, Stowers Johnson in his 1964 travelogue *Gay Bulgaria* claimed that there were no bread lines in Bulgaria, where "there was fresh baked bread each day coming from the central bake house. It was just so good, fresh and hot that people lined up to receive it."[150] Perhaps that was true in Sofia, but for the first time, Bulgaria, like other satellite states (including Hungary and Czechoslovakia) and the USSR itself, began to negotiate grain imports with the United States, Canada, and Australia, securing the shipment of hundreds of thousands of tons.[151]

East-West trade had commenced in the wake of Stalin's death, developing gradually thereafter, but for the United States, trade with the Bloc—especially in wheat—was extremely controversial. On the one hand, a number of US policy makers and constituencies were in favor of wheat exports as a way to gain diplomatic leverage over Eastern Europe, reduce those countries' dependency on the Soviet Union, and dump surplus grain to the benefit of American farmers.[152] President Kennedy famously justified the sale of fifteen million tons of grain to the Bloc in 1963 as a "move towards peace" and a way to "advertise to the world, as nothing else could, the success of free agriculture."[153] But trade in grains to the Eastern Bloc was met with criticism within the United States as a violation of congressional legislation against selling subsidized farm products to communist nations. As US wheat was granted large subsidies as a matter of course, wheat prices were, in effect, kept artificially low.[154] This may have been acceptable in the context of the "food for peace" program, but it seemed suspect to prop up "enemy" regimes. Still, as Kennedy argued, the Bloc could also be seen as just another market or customer for American farmers. More importantly and portentously, Kennedy pointed out that the Russians and other Bloc partners would have to use their precious gold and currency reserves to purchase grain—money that would then be diverted from military use.[155] Indeed, import needs set off a cycle of Eastern Bloc dependency on imports of Western grain, which was in part responsible for growing debt across the region.

In practice, the efforts to produce grain to feed animals instead of people meant that deficits in wheat and imports from the West continued.[156] In 1972, wheat deficits were so severe for the Bloc that the Soviets clandestinely bought up about one-fourth of all US grain (ten million tons) through private brokers, causing prices to spiral upward. What came to be known as the Soviet "Great Grain Robbery" of 1972 caused global food prices to spike by 50 percent.[157] It also eventually provoked the development of a world food security plan that entailed the coordination of an international grain reserve.[158] This rise in imports to meet the domestic demand continued in the 1970s, increasing to twenty-eight million tons by 1979. The implications within the Bloc were dire. As the USSR could no longer provide enough grain for itself, let alone its Bloc partners, the latter turned west and were pulled ever deeper into ultimately fatal fiscal dependencies in the course of the 1970s.[159] Debt mounted across the Bloc, and by 1979 Poland, for example, owed $15 billion. Bulgaria was in slightly better shape, but its $3 billion debt was the equivalent of three times its annual trade with the West.[160]

This import crisis, mind you, was not the result of the inability to produce enough wheat. Instead, it was the result of Bulgarian and Bloc efforts to shift grain production to fodder and diets away from bread. In Bloc-wide planning, carried out through the Comecon, the message was clear: socialist citizens had more than enough bread. At a 1972 Comecon meeting in Budapest, for example, the Standing Commission on Food Industries circulated a set of statistics, plans for food consumption across the Bloc. According to the 1972 numbers, Bulgaria was consuming 195.4 kilograms of "bread products a year," lower than the 1963 numbers but still by far the highest in the Bloc after Czechoslovakia with 156 kilograms. The Comecon plan called for Bulgaria to reduce consumption of "bread products" to 93 kilograms per capita by 1985, with all Bloc countries slated for a decrease, except for East Germany (which was already at 94.9 kilograms per capita).[161] Inter-Bloc economic planning, then, was driven by agreed-upon nutritional norms. And the "plan," so to speak, was to push bread consumption down, while meat, milk, vegetables and fruits, and other foods were slated for increases.

In spite of such plans, Bulgarian citizens continued to savor their white bread and *banitsa*, driving fresh concerns among nutritional scientists. By the 1970s, a trickle of critiques on the negative effects of industrialization of food focused on the ill effects of the shift to white bread in Bulgarian cities and rural areas. Tasho Tashev, for example, noted that the communist period had brought enormous improvement in the quantity and quality of bread available. He lamented, however, the fact that all Bulgarians now "preferred white bread," which was lacking in nutrients, including protein. As a result, he called for "propaganda" to encourage the consumption of whole grain bread.[162] As it turned out, the successful industrialization of bread production had contributed to nutritional problems in Bulgaria. Other nutritional scientists echoed Tashev with discussions of the overconsumption of bread and pastries that were laden with sugar and other carbohydrates. They recognized that the capitalist world was facing a new epidemic of obesity caused by "irrational consumption." Even in socialist Bulgaria, they admitted, as much as 30 percent of the population was "overeating," and carbohydrates were the main culprit.[163] But now that white bread was held in such high esteem, a switch back to "peasant" bread could be a hard sell.

Still, a backlash against industrialized food—including bread—became a subtle narrative thread in a variety of food writings, from cookbooks to how-to books. Such tendencies were by no means just Cold War echoes of the "hippie food" movement that arose in the same period in the West.[164] The timing was similar, but the impetus and forms were local and strikingly different. By the mid-1970s, they found inspiration and support at the highest levels of the establishment, including the influential daughter of Bulgarian leader Todor Zhivkov. Liudmila Zhivkova, discussed in depth in the next chapter, was a self-proclaimed vegetarian, who among other things supported the 1977 republication of the 1926 three-volume *Bŭlgarska narodna meditsina* (Bulgarian folk medicine) by Petŭr Dimkov.[165] The foreword by Stefan Todorov praised the volumes' value as "folklore" but also touted their "factual information."[166] Dimkov himself was an avowed proponent of natural foods as medicine, advocating a plant-based diet, which included whole grain bread. Bread, he argued, was a "fundamental" food, even central to the diet, but he also advised that as little white bread as possible should be consumed.[167] Dimkov invited Bulgarians to reconsider "peasant" modes of eating as a kind of national treasure, but also as a "rational" approach to consumption. The publication of his collected works coincided with the general tendency toward a heightened and overt use of nationalism under the Zhivkovs, which allowed for a revisiting of peasant practices as an important repository of national wisdom. Peasant rationality, or austerity, offered an alternative of sorts to the consumerist tendencies of the "good life," which were pushing socialism into an abyss of debt, overconsumption, and unfulfilled desires.

This rethinking of the official stance on peasant food can also be seen in the first avowedly *national* cookbook of the socialist period. Liuben Petrov's 1978 *Bŭlgarska natsionalna kukhnia* (Bulgarian national cuisine) provided a detailed lineage for the various influences, ingredients, and thousands of dishes now officially coded as "Bulgarian." The goal was to "revive" the "rational heritage" of Bulgarian traditional foods and integrate them into home cooking, but also into communal or public food offerings, like cafeterias and restaurants.[168] As in other official sources, the book lamented many of the problems of the bread-based diet, both in the past (malnutrition) and in the present (overeating).[169] At the same time, the book offers fifty detailed pages of bread and grain-based recipes—not plain old white bread, but a variety of pitas and *banitsas*, and grain dishes like *kachamak* (polenta) and *popara*, some of which featured whole grain flours. If in the past baking bread had been seen as irrational and a waste of time, something from which women needed to be liberated, by the 1970s it was recast as an acceptable and productive use of leisure time. Socialist women, in fact, were called on in a wide range of sources to manage their families' budgets and bodies—to bring them golden, flaky pleasure, but somehow without fattening them up. In the how-to household encyclopedia *Kniga za vseki den i vseki dom* (The book for every day and every home), women were specifically asked to limit baked goods and sugar.[170] Breads became synonymous with both "productive" socialist leisure and a wholesome *national* tradition, but also with the potential for "fattening" and excess.[171] Still, baking bread fit well with the new notions of a do-it-yourself culture that socialism increasingly encouraged, associated with values of productivity and thrift. The utopian future was still colored by consumer promise, but it was still tempered by notions of socialist restraint, prudence, and efficiency. Such values helped shape and delimit, if not justify, socialist consumption, but they were also essential, given economic realities. Thrift and DIY were becoming even more of a necessity in the face of shortages and spotty distribution and the reality of ballooning national debt.

The global recession of the mid-1970s and the oil crisis of 1979 hit the Bloc hard, given its growing dependency on Western trade and capital. Consumer expectations, which had been raised in the 1970s, were dashed by the

FIGURE 1.3. Grocery store in the TsUM Department Store in Sofia, circa 1970s. http://www.lost bulgaria.com/?p=3224.

serious fiscal slump as crippling debts now came due. As shelves rapidly emptied, there was a serious crisis of legitimacy across the Bloc that provoked ambitious reform efforts such as the economic restructuring (*perestroika*) and openness (*glasnost*) spearheaded by Soviet general secretary Mikhail Gorbachev. But not all Eastern Bloc leaders, Bulgaria's included, were open to serious reform. At the same time, state-supported narratives on the successes of the system never ceased. Pŭrvan Todorov's *Golemiiat khliab* (Big bread), for example, boasted, "The big bread that at one time we could only imagine in fairy tales and that people dreamed about—has come! . . . The seed is sown by machines. The earth is plowed by machines. The harvest and threshing—with machines." This abundance of bread, he pointed out, had come as a result of mechanization, "not 'prayers, vows, and animal sacrifice' as in the past."[172] There was some truth sandwiched in between the layers of such propaganda, but by the 1980s it also began to ring hollow. Bad harvests meant that Bulgaria continued to import grains intermittently, as in 1985, which meant mounting debt.[173]

By 1989, the collapse of communism in Poland and Hungary had created a domino effect, which made its way to Bulgaria by November. Mass protests brought the system to its knees, and free elections were scheduled for 1990. Far from offering an immediate solution to all the system's problems, the

FIGURE 1.4. Bread line in Sofia in 1991; the food crisis after the collapse of communism. http://www.lostbulgaria.com/?p=1169.

collapse of communism brought an immediate and dire economic crisis that lasted for over a decade. Ironically, it meant at least a partial return to a diet reliant on bread, with meat and milk production and consumption dropping dramatically.[174] This reversion to bread was also coupled with an overall drop in food consumption in comparison to the communist period. With food subsidies a thing of the past, and with the local food industry in free fall, the purchase of food now required a far greater proportion of the average household income. As a result, from 1990 to 1997, consumption of baked goods decreased by 14 percent, meat products by 33.5 percent, and milk products by 50 percent.[175] During the so-called bread crisis of 1996–97, production dropped by roughly two-thirds, as prices for bread skyrocketed, sending the message that capitalism would not cure all ills.[176] In fact, nostalgia for the cheap, fresh-baked bread of the communist era was awakened by the death of the state-run bakery under post-socialism and the appearance on an unprecedented scale of processed, pre-sliced, and wrapped white bread. As of 2016, Bulgaria was still the highest consumer of bread in the European Union, at 131 kilograms per person per year. This was well below the 1962 figure of 195.4 kilograms but still much higher than bread consumption in the rest of the EU, with Germany and Austria the next closest at 80 kilograms and Spain on the low end at under 50 kilograms.[177] Bread, in this case, could be seen as a symptom of continued Bulgarian "backwardness" in relation to Europe. But perhaps it also signified a deeply embedded love of the essential loaf, in whatever form.

There is no food that is more important in Bulgarian history than bread. Bread has long been a source of daily sustenance for the bulk of the population, and also an important factor in national politics and geopolitics. Lack of bread was the spark for revolution in 1918, and bread was central to the biopolitics of the postwar communist regime. Under socialism, bread mutated from largely unleavened whole grain flatbread to fluffy white factory-produced loaves, and was central to radical changes in postwar production and consumption. Leavened bread and pastries like *banitsa* went from holiday fare or rarer consumption to a regular part of many diets. But for all the system's apparent success in securing bread as part of the socialist good life, bread also became something of a hindrance to dietary and fiscal objectives. The bread-based diet remained a marker of dietary backwardness in relation to the West, and newly refined bread products threw off targets for "rational" consumption. Bulgaria and the Bloc attempted in vain to reduce bread consumption, but instead production shortfalls necessitated costly imports of grain and Western debt. In the end, even state-sponsored publications began to question not just the carb-heavy bread diet, but the shift to refined white bread. Still, Bulgaria rather fortuitously avoided at least some of the pitfalls of industrialized bread—and obesity—experienced by the United States. During the Cold War, it never achieved the American-style loaf: pre-sliced, plastic-wrapped Wonder Bread, devoid of natural nutrients, fortified, and with an impressive (but disturbing) shelf life. Bread marked with this dubious sign of "progress" came only after socialism.

CHAPTER 2

Vegetarian Visions and Meatopias

Morality, Pleasure, and the Power of Protein

Todor Zhivkov, the dictator of socialist Bulgaria from 1956 to 1989, was an avid hunter—so much so that his personal chef for over two decades, Iordan Stoĭchkov, later published a cookbook capitalizing on his culinary experiences with Zhivkov on the hunt, titled *Lovnite retsepti na Bai Dancho: 100 kulinari idei za nai-vkusen divech* (Hunting recipes of Bai Dancho: 100 culinary ideas for the tastiest wild game). Promoted through television and web-media appearances, Stoĭchkov's recipes and musings struck a chord of nostalgia with Bulgarians hungry for intimate revelations from the Zhivkovs' digestive and leisure realm. Killing big game was a popular pastime for Cold War Communist Party elites; Leonid Brezhnev, Nicolae Ceaușescu, Fidel Castro, and others routinely accompanied Zhivkov on his hunting trips.[1] An integral part of the experience was grilling, roasting, frying, or otherwise preparing the product of the day's hunt. Stoĭchkov's book fondly details his mouthwatering preparations, from *sarma* (cabbage rolls) with wild pig to freshly grilled rabbit *kebab* or deer's heart. One can almost smell the roasting meat and imagine the dictators of late socialism gathered around a crackling fire after a day's hunt, recounting their conquests of the day while they feasted on the spoils.

Zhivkov hunted not just in Bulgaria, but in virtually every foreign country he visited, a luxury that required elaborate security arrangements. According to the memoirs of one of his bodyguards, his security detail would often corral wild animals and herd them in his direction, to ensure and expedite his success on the hunt.[2] His achievements as a predator of other animals, then, were at least in part choreographed, though not to the same degree as those of Ceaușescu, his dictatorial neighbor to the north and hunting partner, whose macabre hunting habits—which included mowing down bears with a machine gun—were supported by staff who located, fattened, baited, and herded bears for his killing pleasure. Ceaușescu was reported to have killed twenty-four bears in one day in 1983—with the help of hundreds of forestry and security men.[3] Ironically, Ceaușescu became so hated by his own people that he himself was hunted down and shot as the Romanian socialist system collapsed in 1989. This was not the fate of Zhivkov, whose image as a hunter was tempered by his relative restraint and folksy "man of the people" persona. Notably, according to Stoĭchkov, Zhivkov "never ate his prey," even as his many distinguished guests were always left "licking their fingers" after gorging on the chef's tasty preparations.[4] Big-game hunting was not only about the meat; it was also about power, pleasure, and affluence—the limited and exclusive access to the resources needed to hunt and kill big game, which had long been the purview of elites. State socialism created a system in which new elites like Zhivkov enjoyed this privilege.

But it was also a period in which meat was more widely available, affordable, and avidly consumed than at any other time in Bulgarian history. Transformations within agriculture, animal husbandry, and meat processing increased that availability, and meat, especially pork, became accessible to the masses on an unprecedented scale. Newly engineered modern pig breeds replaced "wild" pigs in the Bulgarian diet so that the communist "new man" could now be fortified with the meat of the "new pig." An increase in supply was accompanied by a change in meat-eating culture, as new nutritional, culinary, and religious norms enabled a gradual transition from a diet in which half of the year was meatless to almost daily meat for all. In the past, meat consumption had been limited for much of the population by financial considerations, and for Bulgaria's Orthodox Christians by the religious calendar with its frequent meatless fast days. Under socialism, such religious strictures were deemed irrational and even harmful to the goal of building socialism. The modern socialist worker or collective farmer required a more consistent, and now possible, animal protein–filled diet. The flavor and scent of meat became part of everyday life for a large segment of the Bulgarian population under communism, including the grill-sizzling, lush fattiness of the popular seasoned *kebabche* (elongated grilled meat patties), *kiufte* (grilled or fried meatballs), or *kebab* (grilled meat in chunks, as in shish kebab), or a thick, hearty meat *giuvech* (stew). Whether boosting work or flavoring play, meat was a primary ingredient in building socialism and, by extension, making Bulgaria modern in the postwar period.

Bulgaria's carnivorous turn was shaped by local conditions and narratives, but it also echoed the Bloc-wide and global "nutritional transition" of the modern period. Meat had long been associated with power and access, abundance and wealth, and its eventual democratization was first tied to shifts in consumer culture and changes in nutritional science in the West. In the United States and parts of Western Europe, for example, a range of modern meat-producing practices and technologies, including the mass slaughterhouse and refrigerated transport systems, led to a dramatic rise in meat consumption by the late nineteenth century.[5] By the twentieth century, Western modernity smelled of meat, and its seeming power was increasingly undergirded by the new gospel of food science. Meat's fat and calories, and later its protein, were seen as the best antidote to hunger and weakness. The real and imagined properties of meat became connected to the imperative of building well-fortified modern societies. Meat was marinated in powerful symbolism, and as it fortified ever greater segments of the population, it became an indicator, and for some the driving force, of affluence, abundance, and progress.[6]

Communist thinkers and leaders were well aware that the increased production and consumption of meat had accompanied—if not driven—industrialization in the West. They wanted to follow this sinewy path in a veritable "meat race," but without the accompanying exploitation of workers, peasants, or weaker states. Instead the communist world looked to build its own "meatopia," using the power of protein to fuel the transformation of socialist bodies—and by extension socialist society. In the communist bioimaginary, meat had the potential to both drive and confirm socialist progress via a high intake of calories and protein, while providing savory substance to the socialist good life.[7] In the Bulgarian case, the ambition to ensure a steady diet of meat also tapped into longtime local (and foreign) observations about the "backward" Balkan peasants' diet, which was emblematic of their geographic and symbolic place on the edge of Europe.[8] In a sense, Bulgarians' fragile status as semi-European required a beefing up of their diet, and the communist state put considerable effort into serving up this kind of a future. With meat consumption on an upward trajectory for the whole period, the 1980s witnessed the highest rates of meat production and consumption in Bulgarian history.

At the same time, there was room at the table for meatless alternatives to such a future under late socialism even at the highest levels of power. As noted in the last chapter, vegetarianism was not new to Bulgarian history. This history provided Todor Zhivkov's influential daughter, Liudmila Zhivkova, scaffolding for her personal embrace of vegetarianism in the 1970s. Her support of other vegetarian voices constituted a serious challenge to the meat-eating socialist establishment, especially as she deployed the language and rationale of the system in *rejecting* meat. Her challenge notwithstanding, meat provided communists with a kind of connective gristle in their scramble for legitimacy in the tumultuous 1980s; the scintillating flavor and caloric heft of meat was of critical importance, tied to progress, pleasure, and even national identity. But perhaps equally important was the system's ability to allow room for vegetarian dissent from the meat-eating norm, even after Zhivkova's untimely death in 1981.

Meat and Power

The history of meat is best understood against the backdrop of humanity's evolving relationship with animals—as friend, foe, and food. For most societies at some point in history, hunting, and eventually domesticating, animals became critical to survival. As a food source that required skill and tools to kill, slaughter, and cook, meat fueled human ingenuity and fed the gradually expanding human brain.[9] As some scholars have argued, it was the means and modes of meat procurement and consumption that, over a period of two hundred thousand years, made us "human." Meat literally transformed our brain capacity and physiology, while patterns of meat consumption played a role in encoding power relations—man over animal, and man over man or woman.[10] Meat was habit forming, a ready source of protein and B vitamins as well as gastronomic pleasure. And yet, as scholars have rightly pointed out, the human body—in particular our teeth and stomach—is poorly designed for chewing and digesting raw meat. It was only with fire, the most important of human tools, that cooking made meat more edible. It was not only *eating* meat, then, that arguably made us human, but our ability to *cook* it and hence make it digestible, palatable, and pleasurable.[11]

As meat biochemically transformed bodies, it also entered the bioimaginary of various cultures in disparate ways. Virtually every society has been compelled to decide whether, when, and how animals could be eaten. Since ancient times, prominent thinkers and religious movements have challenged the morality of killing animals and eating their flesh and organs. In many cases, animals were seen as deities with considerable powers. In others, the issue was the bodily pleasure of eating flesh, if not the actual killing of animals, which was subject to regulation, restriction, or prohibition. Many religions dealt with the issue through ritualized animal sacrifice or slaughter, while others were explicitly vegetarian. Throughout history, prominent individuals and groups have questioned the practice of meat consumption as an ethical, healthful, or sustainable practice.[12] At the same time, the notion of humans' dominance over the natural and animal world became increasingly connected to notions of progress and development.[13] Eating ever greater quantities of meat exemplified economic and political power, and in some cases came to be tied to the purported "superiority" of the developed and regularly meat-eating nations of the modern West.[14]

The writings of Western travelers to the Balkans provide a window into the way that Balkan and other "non-Western" practices around food became objects of observation and normative judgment. The tastes and smells of local food and drink and the associated habits and etiquette were woven into travel narratives that were often meant to shock and titillate with the "barbaric" or pleasurable realities of the "Orient." Foreign assessments of the seeming lack of civilization in the Balkans in some cases focused on the diet, but also on the "Oriental" manner of eating meat with the hands, or with the carving done in front of guests, as opposed to the more common Western practice of obscuring the animal carcass.[15] In the Bulgarian case, foreign observers were also bewildered by the continuous feasts and fasts, which many saw as "a strange medley of old Slavonic paganism and the ill-defined and superstitious Christianity which exists in the East."[16] If some Western observers were appalled by the large number of feast days each year, they were just as judgmental about the 182 days of meatless fasting, which "weaken the peasant by their extreme rigour (particularly as they occur only on the days when he may labour), as his diet is reduced to bread, onions, garlic, or one of the few kinds of vegetables which he cultivates."[17] In a sense, these seeming caloric imbalances—and not Ottoman Turkish exploitation— were deemed to be at the root of Bulgarians' lack of productivity, and hence their "backwardness." But not all foreign observers saw the feast and fast— or the lack of consistent meat in the Bulgarian diet—as detrimental to the local constitution. H. C. Barkley's 1877 *Bulgaria before the War*, for example, lauds Bulgarians' simple grain-based diet and their resulting physical rigor.[18] In contrast, Barkley viewed local Turkish elites as gluttonous, describing a "goul-like proceeding" when an "old Pasha" unceremoniously gouged a hole in a roasted sheep's side and "groped about" with his hand until "he secured the heart and hauled it forth."[19] Then again, Barkley admitted that some of the best dishes he had ever eaten were in Turkish houses, like the "candied chicken" he had eaten for breakfast while a guest of the pasha of Varna.[20]

Unlike local pashas, and like most Christian peasants under Ottoman rule, the bulk of Bulgarians had meat-poor diets, due to religious practice, practicality, and generalized poverty. As noted above, Orthodox Christianity had a large number of meatless (and dairy-free) fast days, including every Wednesday and Friday and the forty days leading up to Easter.[21] Adhering to the meatless fast was a strong indicator of Orthodox faith in the region, as it differentiated Christians from Muslims and Jews (whose fasting practices were not specifically meatless).[22] Not all Bulgarians strictly observed the fast, as indicated in the common saying "Doide gost, razvali post" (A guest comes and the fast is broken).[23] But members of the Orthodox community, particularly monks, proved their piety through fasting—especially forgoing meat, which was associated with "pleasures of the flesh."[24] The most famous saint in Bulgarian history, the monk Ivan Rilski, was well known for his completely meatless diet and severe fasting.[25] The meatless fast was also interwoven with pre-Christian pagan beliefs that had persisted in local folk practice, in which forgoing meat was regarded as a healthful practice that went beyond pious concerns.[26] Finally, historically there were sound economic reasons to limit the consumption of meat and other animal products in rural Bulgaria. Animals were needed in the rural economy for subsistence or to build wealth, and killing animals meant losing sustainable sources of milk, wool, breeding stock, labor, transport, or trade.

Fast days and periods alternated with non-fast days and "feasts," which often centered on ritual slaughter or *kurban*—the Turco-Arabic word for "sacrifice." The feast was at heart an extension of the local pagan tradition of animal sacrifice; in Islam, Judaism, and Christianity, animal slaughter was often tied to ritual ceremony and celebration. Orthodox Christians regularly feasted on roasted lamb or sheep to celebrate saints' days, including Saint Georgi's Day (May 6), and other holidays or celebrations. The main exception was Christmas, when it was customary to slaughter and roast a pig.[27] Leftover meat and organs from such feasts could then be used to create traditional dried and cured spiced meats.[28] These, as well as preserved pig fat (or lard), could be rationed and eaten over much of the year, or sold in nearby towns (to Christian dwellers) for export to more distant Ottoman urban markets.[29]

With the Ottoman conquest in the fourteenth century, the Bulgarian-populated territories of the eastern Balkans became a critical part of the food supply line to the expanding Ottoman imperial machine, especially the capital city, Istanbul. The region's geography was well suited for raising, breeding, and pasturing animals, particularly sheep and goats, given its fertile summer pastures in the mountains and its lowlands near the warm Black and Aegean Seas. Larger livestock, such as cows, oxen, and water buffalo, were most often found in the valleys and plains of the region, associated with the agricultural estates of Ottoman Turkish elites.[30] In the mountain villages where Bulgarians predominantly lived, some larger stock were present for labor and transport, but sheep and goat herds were better able to navigate the terrain. In the seventeenth through the nineteenth centuries, a steep spike in Ottoman demand for Bulgarian animal products—meat and dairy products as well as wool, tallow, and leather—reflected the growing needs of the empire.[31] In 1674 alone, the territories of Ottoman Bulgaria supplied some seven million head of sheep to the imperial capital, including ten thousand for the sultan's kitchen. The Bulgarian National Revival (late eighteenth century to 1878) is largely associated with the new wealth of wool merchants from the Central Balkan Mountains. But Bulgarian shepherds and merchants also became critical sources of animals for meat for Ottoman cities, as well as for the armies, which were essentially cities on the move.[32] The Strandzha region in southeastern Bulgaria was the most important in terms of meat supply for the gargantuan market of Istanbul. It was so close that animals could be transported on the hoof—a critical factor in this period when refrigerated transport was unavailable.[33] But Ottoman armies also marched across Bulgaria or occupied various garrisons for long periods, drawing on local meat supplies as they went.

Local populations had a wide variety of meat sources—pigs, chicken, wild game, and seafood—but sheep were the most widely domesticated and consumed animals in Ottoman Bulgaria. As Bulgarian merchants entered the meat trade and managed huge herds, they also became avid consumers of mutton and lamb on the pattern of the urban Ottoman elites who provided their primary culinary inspiration. Pigs were far more popular in the rest of Europe and throughout the Slavic world, including the adjacent Serbian and Romanian lands. But as pork consumption was prohibited by Islam, pig breeding was discouraged, if not wholly absent, in Ottoman cities, and the slaughter or even foraging of pigs was subject to special taxation in the region as a whole for much of the period.[34] In Ottoman Bulgaria, as a core territory of the Ottoman Empire, pork tended to be a rarer treat for Christians, with at least one hog a year fattened, slaughtered, and consumed for Christmas, its fat and leftover meat smoked or otherwise preserved for later use.[35] Whatever the form, meat was eaten in spare amounts by peasants, the bulk of the local population.

But the same was not true of the rising Bulgarian elites of the late eighteenth and the nineteenth century.[36] For many who grew up in smaller towns and moved to Ottoman cities, the tantalizing sights, smells, and tastes of Ottoman street food or café and tavern fare must have been a sensory revelation. With the variety of markets, shops, street stalls, and small establishments, meat could be consumed every day, provided that one had sufficient funds. As of the nineteenth century, Ottoman cities had ample guilds for butchers and more centralized butchering stations, but also *kebabchi* (kebab cooks), who were part of the fabric of Ottoman foodways.[37] As noted in the last chapter, Istanbul-based Petko Slaveikov, a renowned figure in the Bulgarian National Revival, was intent on spreading the elaborate culinary culture of the Ottoman capital, itself a smorgasbord of influence, to the new Bulgarian elites of the provinces. Notably, pork was practically absent from his 1870 *Gotvarska kniga*, the first Bulgarian-language cookbook. The recipes there for *kebab* and *kiufte* primarily called for lamb, with some veal, chicken, and rabbit.[38] Only in the latter portion of the book did pork appear, in a section on "foreign" methods of preserving meat by salting it and forming it into salamis, which was not yet a common practice in the Ottoman Empire.[39] It was far more common in the core Ottoman lands to dry and cure spiced meats, most commonly as *sudzhuk* or *lukanka* (from ground meat) or *pasturma* (from whole pressed meat).[40] Interestingly, Slaveikov included only a small section on *sudzhuk*, in favor of a much larger section on salting and curing salamis.[41] Still, the European techniques he integrated into his cookbook did not eclipse the flavor and spice of Istanbul-influenced dishes, which were still front and center.

But just as new elites like Slaveikov came to partake of and spread Ottoman elite culture—culinary and otherwise—more radical Bulgarian voices disparaged the implications of such practices. Bulgarian revolutionaries, such as the poet and publicist Khristo Botev, were critical of Ottoman oppression and all those who maintained and supported it. This included many Bulgarian merchant and administrative elites, the so-called *chorbadzhi*, who benefited from Ottoman and European power and trade. While the Turkish term *chorba*, derived from Arabic, is used in Bulgarian as a general term for soup, historically it was associated with a rich meat stew that would have been regularly consumed by local elites or *chorbadzhi* (literally keepers, or eaters, of the meat soup).[42] Most revival elites were children of *chorbadzhi*, often only a generation or two removed from their peasant origins. For many, their recent social mobility only heightened their awareness of the gap between peasant subsistence and the fattier, meatier diet of the *chorbadzhi* and other Ottoman elites.

In many respects, the extrication of Bulgaria from Ottoman rule only widened this socially attenuated dietary gap. True, in the wake of the 1877–78 Russo-Turkish War, which secured Bulgaria's autonomy, larger Turkish landholdings were handed out to landless Bulgarian peasants.[43] But the war and its aftermath were also devastating to the economy of the emergent

Bulgarian principality, especially its economic base in animal husbandry. New Bulgarian borders and tariff zones cut off local herders from the large Ottoman markets of the past, while creating problems for summer and winter pasturing and transhumance herding. This meant that many decades would pass before there was even a partial recovery in the local meat economy. Meat exports did slowly expand as trade with the Ottoman Empire resumed, accompanied by a robust reorientation toward Central European markets. The increased presence of Central European and Russian émigrés in Bulgarian cities contributed to a slow shift toward increased consumption of pork, alongside the traditional sheep-based meat economy.[44] Meat remained a common component of the diet for the affluent in the cities, but it was still a rarity for the new urban and rural poor. In fact, with growing poverty, land parcelization, and cycles of peasant debt in rural Bulgaria in this period, meat was likely consumed even more rarely than in the past. Moreover, as poverty drove many peasants to seasonal work in Bulgarian cities, their diets deteriorated owing to their less frequent access to the family garden and cellar. For many, these worsening economic ills generated deep disillusionment with the post-Ottoman order and fueled the popular refrain "Ot tursko po-losho" (It is worse than Turkish times).[45]

A kind of nostalgia for particular elements of the Ottoman past was not limited to the peasantry. As Bulgaria's new urbanites assumed their place as political-administrative, economic, and cultural elites, they embraced and expanded Ottoman food culture, which included the grilled meats that were readily available in a range of local food establishments or at street food vendors.[46] They also became avid consumers of the dried prepared meat products—*sudzhuk*, *lukanka*, and *pastruma*—that hung in shop windows and were sold in market stalls, especially at the new *Khali*, or covered market, completed in Sofia's city center in 1911.[47] New "European" (especially Central European) techniques and preparations for curing and processing meat came into vogue, along with accompanying drinks. But Ottoman forms also had amazing staying power within a new urban food culture. *Skara*, or grilled meats (especially *kebabche*), became a culinary mainstay in the post-Ottoman Bulgarian city. But so too did urbanites embrace the tradition of the family "picnic"—almost always featuring meat, prepared in advance or grilled in the open air—in parks, or as part of pilgrimages to distant sites of natural beauty, including monasteries.[48] Ivan Vazov, for example, in his 1894 novel *Under the Yoke*, expressed a marked nostalgia for the flavors of simpler times.[49] For Vazov, the pleasures of local food and drink offered a means for national survival under the Ottomans: "A nation, however hopeless its bondage, never ends its own existence: it eats, drinks, begets

FIGURE 2.1. Bulgarian butcher shop, circa 1930s. http://www.lostbulgaria.com/?p=2120.

children. It enjoys itself. If one but look[s] at the poetry of a nation, one finds clearly expressed the national spirit, the nation's life, and its views of existence. There, amid cruel torments, heavy chains, dark dungeons, and festering wounds, is yet interwoven the mention of fat, roasted lambs, jars of red wine, potent raki, [and] interminable marriage feasts."[50] It is not that Vazov called for a return to a time of "heavy chains," but rather he set the tone for a generation of Bulgarian writers, whose prose looked for inspiration to Ottoman and post-Ottoman Bulgarian everyday life—including the pleasures of food and drink.

Vazov was also the most prominent figure in Bulgaria's back-to-nature movement, which dovetailed with late nineteenth-century movements in the West. Four years after publishing his iconic first Bulgarian novel, Vazov penned his magisterial travelogue *The Great Rila Wilderness* (1892). In it he details his quest to leave behind the deleterious effects of "civilization" in order to experience the Bulgarian wilderness and its earthy pleasures. As Vazov described it, he felt at home with the shepherds, his guide, local peasants, even a brigand—unspoiled Bulgarians who lived far from "our sick civilization."[51] For Vazov, escaping from civilization was more than just a religious or spiritual endeavor; it was also a *national* activity. As a curative, he implored Bulgarians not only to explore the vast natural beauty around them but also to connect to the simple local forms of consuming, framed not just by lamb on a spit, but by homegrown traditions of restraint. Part of Vazov's journey was a pilgrimage to the Rila Monastery and the nearby cave where the renowned monk Ivan Rilski lived out much of his life in the ninth and tenth centuries. Known for his extreme piety and fasting—the hallmark of his saintly existence—Rilski, as Vazov noted, "lived exclusively on plants and ate only enough to keep his body and soul together."[52] Incidentally, Vazov packed a side of bacon provided by local monks when he journeyed to Rilski's cave.[53] Vazov waxed poetic about both Rilski and even older traditions of restraint that had deep roots in Bulgarian soil. Namely, he called on the memory of Orpheus, the poet and prophet in Thracian and Greek mythology who was an animal charmer and opposed the consumption of flesh.[54] If he did not personally follow this meatless path, he nevertheless conjured the spirit of the ancestors as a model of purity and piety to contrast to the decadence of the modern age.

Vazov's "return" to nature and the Bulgarian past was part of a growing global critique of the ill effects of "progress" that gained momentum by the turn of the century. For a growing number of influential thinkers, this critique turned directly against meat eating in favor of the ethical and curative powers of vegetarianism. A number of homegrown Bulgarian vegetarian movements appeared in the early twentieth century, which looked outside and within for global and local inspiration. A major figure in this regard was Lev Tolstoy, whose writings and practices offered a ready non-Western, Slavic model for an ethical, meatless way of life. His legendary vegetarian treatise "The First Step," published in 1892 as a preface to the Russian edition of Howard Williams's *The Ethics of Diet*, made him an influential voice in a global vegetarian movement that advocated a "return to the soil" and a selective embrace of peasant modes of producing food. Overtly critical of gluttonous meat-eating elites, Tolstoy tapped into the sparse Russian peasant diet—or at least its meatless fast—as a way of life.[55] His notion of "returning to nature" by adhering to a peasant way of life on his own estate was in part sparked by his repugnance toward the methods of meat processing enabled by slaughterhouses, which in Russia, as in the West, had become industrialized killing centers.[56]

The global development of modern meat production techniques arguably furthered the expansion of an international vegetarian movement. The global epicenter of modern meat production was Chicago, where stockyards and slaughterhouses supplied the large urban populations in the American Northeast.[57] Interestingly, the Third International Vegetarian Congress was held in Chicago in 1893, the same year as the famous Chicago World's Fair.[58] The congress attracted far less attention than the Chicago slaughterhouses, which were a popular tourist attraction for visitors to the fair, as part of the spectacle of American progress. Significantly, one of Bulgaria's most famous authors, Aleko Konstantinov, described his own trip to the Chicago Fair in his celebrated 1893 travelogue *To Chicago and Back*.[59] In one of the book's earlier episodes, he dines at the "Turkish" restaurant on the fair's famous Midway, where "an American with a fez" tends "a smoky grill piled with fresh kebabs."[60] The allure of Turkish cuisine was no surprise, given his disgust with American food: "There could be no place on Earth a less tasty cuisine than the English, as implanted in America," with its "bloody unsalted meats with pale sauces."[61] But his ability to hold down this meal was apparently imperiled by his side trip to one of the famed Chicago slaughterhouses. As he stood there surrounded by the putrid and overpowering smell of pig death, his admiration for American material progress was soundly shaken. In an unforgettable scene, Konstantinov almost throws up several times and nearly loses consciousness. His friends continue on the tour while he chokes on the "murderous stench" until he finally yells, "I'm dying!" His physical and moral revulsion are palpable as he ruminates, "What has brought us to this place?"[62] Konstantinov's disgust at Anglo-American animal slaughter and meat preparation was not enough to make him reject meat altogether. Still, his descriptions of the Chicago slaughterhouse are evocative of the various interwoven strands of vegetarian thought that were coalescing in turn-of-the-century Bulgaria.

The Aftermath of Slaughter

Vegetarian movements blossomed in the wake of World War I. The Great War was a catastrophic and transformative global event that deepened already brewing questions about Western "progress" and unleashed a global search for radical solutions to political and social ills. In Bulgaria, the death toll in the Balkan Wars and World War I, postwar hunger, and the heavy weight of defeat caused a radicalization of the masses and a serious existential crisis among elites. As the war came to a close, the Bulgarian Agrarian National Union was swept into power, and the Bulgarian Communist Party emerged as a formidable political force. Both movements espoused a kind of ennoblement of the laboring classes, models of collective restraint in terms of consumption, and opposition to exploitative overconsuming elites. But by 1923, these political moments were pushed to the political margins, if not abroad and underground. Less political movements with alternative social models, however, continued to survive and proliferate. Among them were the two movements introduced in chapter 1 whose ambivalence toward modernity and violence made vegetarianism a central tenet.

For one of those groups, the White Brotherhood, or Dŭnovtsi, vegetarianism had clear spiritual underpinnings. Their leader, Petŭr Dŭnov, had been educated in American Protestant schools, but then had turned to a theosophical spirituality that looked to various world religions, and occultism, for inspiration. But the Dŭnovtsi were also connected to a variant of Christianity from medieval times, Bogomilism. The Bogomils were members of a Christian neo-Gnostic or dualist sect that was founded in tenth-century Bulgaria by Pop Bogomil (Priest Bogomil), and whose influence expanded across Europe over the centuries that followed. Embedded in their practices was an ascetic renunciation of pleasures of the flesh, including meat and alcohol. The Bogomils tapped into practices of monastic bodily restraint, which they believed were not just for the holy few, but for the community at large. This was tied to their search for a direct connection to God, which also called for a rebellion against the established church. Because of this, they have long been referred to, at least by some, as the original "Protestants" of Europe.[63] Indeed, many of Dŭnov's followers believed that their leader was the reincarnation of Pop Bogomil, and many contemporaries referred to the group as the "new Bogomils."[64] Stoian Vatralski, a prominent interwar writer, noted that the White Brotherhood, like the Bogomils, was a source of spiritual renewal in Bulgaria with global significance.[65] This powerful vision of the Dŭnovtsi—their alternative view of the modern world, including vegetarianism and natural healing through food—attracted a large number of followers and sympathizers, including Bulgarian elites.[66]

Another vegetarian movement, Bulgaria's "Tolstoyans"—also introduced in chapter 1—had put down roots in Bulgarian soil prior to World War I. Inspired by the ideas and practices of Tolstoy, the group had expanded in size and their scope of activities by the turn of the century and tried unsuccessfully to found an agricultural commune. In 1907 they finally broke ground on a successful commune in the lush mountain village of Alan Kayrak, nestled in the Strandzha region of the southern Black Sea coast. The village was renamed Yasna Polyana (bright village), after Tolstoy's Russian estate and peasant commune Yasnaya Polyana (with a shortened Bulgarian adjectival ending).[67] Many of its leaders had made a pilgrimage to Russia to visit Yasnaya Polyana before (and after) Tolstoy's death in 1910. Khristo Dosev, for example, spent time in residence there from 1907 to 1909 and became extremely close to Tolstoy and his inner circle. In Dosev's memoir about this experience, he relates his first memorable meeting with Tolstoy, who asked him point-blank, "What do Bulgarian peasants eat?" "Lots of beans," Dosev answered earnestly. Tolstoy reportedly smiled with satisfaction at the answer. Soon after his return to Bulgaria in 1909, Dosev founded the Bulgarian Vegetarian Union, which sought to unite vegetarians and sympathizers of whatever political or ideological stripe and spread the gospel of a meatless way of life.[68]

The Vegetarian Union and the Tolstoyan movement gained in strength after Tolstoy's death, and especially in the aftermath of World War I. The massive human casualties of the war brought even greater urgency to the Tolstoyan project, both in Bulgaria and around the world. For Tolstoyans conceptually tied the carnage of war to humans' treatment of animals. As a 1923 article in the Bulgarian Tolstoyan publication *Vegetarianski pregled* (Vegetarian review) argued, "The war has been over for five years now . . . during which time we have not felt the rumble of bloody battle, but in these five years the war against animals . . . a horrible and bloody war has continued. . . . The killing of animals has not stopped."[69] This kind of continued "war" was condemned as part of a pervasive global culture of violence that was seen as tantamount to self-destruction, if not a kind of symbolic cannibalism: "We commit crimes against animals, we kill them, but in the process we also kill our own souls, we kill ourselves, and we eat ourselves."[70] In response, the movement took action to circulate their ideas and to model a lifestyle that honored life in all its forms. Under the 1918–23 Agrarian regime of Alexander Stamboliski, the Tolstoyans took advantage of state-sponsored opportunities and formed new agricultural "cooperatives" across Bulgaria. Yasna Polyana continued to operate as the headquarters of the movement, while adherents opened a number of other communes, as well as vegetarian cooperative hotels in Bulgarian cities. In Sofia, the first vegetarian restaurant and center for Tolstoyan thought was established in 1919 at the home of the famous writer Todor Vlaikov. It was soon moved to a more substantial space at 3 Sixth of September Street, where it became a kind of discussion club and meeting place for the Vegetarian Union. Its first chef, Stefan Georgiev, trained other cooks, who then cooked at vegetarian restaurants across Bulgaria.[71]

At the same time, the movement's intellectual labor continued, with the translation and production of a large range of publications, including the works of Tolstoy, newspapers like *Vuzrazhdane* (Revival), *Novo obshtestvo* (New society), *Svoboda* (Freedom), and *Vegetarianski pregled* (Vegetarian review), as well as books and pamphlets.[72] These publications had circulations between two thousand and six thousand for much of the interwar period, indicative of the extent of the movement's influence. They reported on the state of the Bulgarian movement but also espoused vegetarianism, nonviolence, and abstinence.[73] Tolstoy's ideas, as articulated in "The First Step," offered a clear philosophical direction, all the more meaningful as it was grounded in a neighboring and "brotherly" Slavic culture—that of the Russian "liberators" of Bulgaria. But Tolstoyan writings also explored the vegetarian ideas of other, more ancient figures and movements of more local origin. As laid out in Dosev's *Etika na khranata, ili Pioneri na vegetarianstvo* (The ethics of food, or, The pioneers of vegetarianism), the rejection of meat on ethical grounds could be traced to the most influential figures among the ancients, most notably Pythagoras, but also Plato, Ovid, Seneca, and Plutarch. This offered the movement a distinguished, if not regional, pedigree.[74] Like the Dŭnovtsi, Dosev also looked to the vegetarian Bogomils for grounding. The Bogomils were important for Dosev both as the first Christian vegetarian movement and also because of their Bulgarian roots.[75] Dosev's work in essence made Bulgarians part of the larger genealogy, history, and ethical dialogues of vegetarianism, which also encompassed Buddha and many figures from beyond the European tradition.

While Western minds had an important place in the pantheon, the technologies and culture of *modern* meat production and consumption were a regular target for Tolstoyan critique. Bulgarian Tolstoyans derided modern civilization for turning man into a pleasure-seeking machine that could "swallow muscles and gnaw on bones" of innocent animals.[76] In such writings, the Chicago stockyards were painted as a kind of mass death camp. As one 1927 article in *Vegetarianski pregled* lamented, "In just one world city, Chicago, 54 million animals, cows, lambs, sheep, pigs, and others, are killed every year, with enough blood flowing from them to fill a huge reservoir. This is just in one city."[77] The article continued: "The same crimes are being carried out in every city and village in Bulgaria and in every other country. Collected in one place, this blood would create a huge stream, an unstoppable river of blood, like the Danube or the Volga, which would constantly flood the very foundations of civilization."[78] Clearly the West was not the only culprit in perpetrating mass death, and Bulgaria was not immune to accusations of mass animal killing. As the author noted with irony, the scale of Bulgarian slaughter became especially unbearable on feast days like Christmas, when "just as people remember Christ for his mercy and purity, the carnivorous feast doubles this river of blood."[79] And yet for Bulgarian peasants, the "feast" at least was tempered by meatless fasting—an explicit model for Tolstoyans. The movement inspired by Tolstoy advocated a kind of year-round meatless fast, equated with Christian piety, but also firmly rooted in ambivalence toward Western "progress."

This ambivalence, however, did not preclude a forward-looking vision, increasingly informed by new research in the sciences and social sciences. Far from retreating to the past, Tolstoyan authors continually advocated change, a "new life," which was only possible without "the remains of death in our teeth."[80] Keeping up with the times, the Bulgarian Tolstoyans enlisted new ideas in nutritional science, economics, and ecology in an effort to convince wider audiences.[81] Stefan Andreichin, perhaps the movement's most articulate advocate of vegetarianism, advanced the notion of a meatless diet as more ecologically sustainable in a world with a growing population.[82] He also pointed out that it was a cheaper and more rational way to eat, given the global interwar economic crisis and Bulgaria's generalized rural poverty.[83] In addition to such arguments, regular testimonials by local doctors attested to the "toxic" nature of meat, which purportedly was making many of their patients sick.[84] Doctors' advice was bolstered by a series of articles that drew on new research in nutritional science—albeit selectively—to offer dizzying detail about plant proteins and vitamins in advocating the healthful benefits of a plant-based diet.[85]

In fact, Bulgarian food science came into its own in this period on a decidedly vegetarian foundation. No figure was more important in this process than the prominent Bulgarian public intellectual, biochemist, and pioneer of food science Asen Zlatarov. As explored in the last chapter, Zlatarov was highly concerned about the Bulgarian bread-based diet, which he believed needed far more nutritional variety—first and foremost protein and vitamins. But he advocated a carefully designed vegetarian diet as the healthiest way to provide Bulgarians (and the world) the nutrients they needed. Widely quoted in the Tolstoyan literature on food science, Zlatarov integrated vegetarianism into his writings and lectures at Sofia University.[86] Although he was not an active part of the Tolstoyan community, Zlatarov was a sympathizer and a major proponent of bringing new protein-rich plants, including legumes like soybeans and chickpeas, into Bulgaria for cultivation. In keeping with his ethical and dietary leanings, he famously hosted the visit to Bulgaria of the well-known Indian poet and intellectual Rabindranath Tagore in 1927. Thousands of Bulgarians flocked to see Tagore, whose ideas on nonviolence, vegetarianism, and social justice were in line with Tolstoyan thought and resonated widely among Bulgarians—intellectuals and the masses alike.[87] In a sense, Zlatarov's own dietary ideas, while employing Western science and even a philosophical grounding, looked east as well as locally for inspiration.

Zlatarov lived and wrote in an age of extremes and a brewing culture war that divided Bulgaria—along with the rest of Europe—between the political left and right. While the Dŭnovtsi were apolitical, Tolstoyans were decidedly left-leaning. They were not a communist movement, but they approved of the fact that the Bulgarian Communist Party, like the Agrarians, had been the only openly antiwar party in World War I. By the 1920s, Zlatarov, like many Tolstoyans, was excited about many aspects of the new Soviet experiment as seen and understood at the time. These included the redistribution of wealth and the turn to science and technology to address social ills. For Zlatarov, the diet directly reflected wealth disparity; his own in-depth studies had revealed the stark discrepancies in protein consumption between the Bulgarian masses and elites, and that Bulgaria as a whole was lagging behind the West.[88] He and others were enamored with the Soviet support of science, education, and agricultural communes—including the new commune established at Yasnaya Polyana by Tolstoy's daughter. The notion that the Soviets were taking on poverty and the West in a drive toward progress seemed to overshadow Stalinist crimes, which were still somewhat hidden from view. Within interwar Bulgaria, left-leaning vegetarians like Zlatarov were increasingly suspect, the object of state surveillance. As noted in the last chapter, Zlatarov, after visiting the Soviet Union in 1936, went into exile in Vienna, where he died soon thereafter. His ideas, however, remained very much alive in Bulgaria. They would take on new meaning in the postwar period.

Red Meat?

At the close of World War II, the newly installed Bulgarian communist regime gradually implemented the most far-reaching project of social and economic transformation in the country's history. Both brains and brawn were needed to drive state ambitions for rapid progress. As a result, the communist state became directly involved in reshaping the bodies of Bulgarian citizens from the inside out via protein, vitamins, and minerals. Marxism-Leninism was an ideology and vision for the future firmly grounded in a veneration of science and technology, not spirituality or belief. In practice, of course, commitment to progress, science, and the future still required bold leaps of faith. And so too did belief in food science, itself still malleable in this period. After all, the newly discovered components of food—vitamins, minerals, proteins, and carbohydrates—were too small to see with the naked eye. Even today, people have to be convinced of their existence and impact. At a time when food science itself was new, this was an even greater task than it is now. Food scientists became a new kind of priesthood who espoused a way forward, poised to take the place of religious dietary rules and local folk wisdom. Scientific-nutritional formulas were readily adopted as long-term objectives of the communist regime that would undergird progress and the building of socialism.

Protein, particularly meat, was potent fuel for the communist bioimaginary. More abundant and equitably distributed meat would help Bulgaria overcome domestic inequality. It would not only help the Communist Party win over the masses, but also put Bulgaria firmly on a path to "catching up" to the West. As the West itself was on a path toward ever greater meat consumption in the postwar period, "catching up" inevitably put the Eastern Bloc in a "meat race"—a competition far more costly than the more well-considered arms and space races. This is not something that the late Zlatarov would have approved of. True, his interdisciplinary interwar food science provided a foundation for paving this pathway forward—his precise proof, down to the gram, that the Bulgarian masses, and even Bulgarians as a whole, were protein deficient. His remedy, however, was not a meat-oriented diet, but rather a more economically viable, sustainable, and in his estimation healthful vegetarian way of life. But Zlatarov was dead and gone, and his ideas could be edited for selective use at the whim of the new regime. Food science was malleable; it could be molded by political expediency or economic necessity, or by belief and imagination, as much as by data.

In 1947, communist state officials organized a congress of local administrators, doctors, and scientists on the topic of "feeding the Bulgarian nation." Citing the past studies of Zlatarov and others, the congress focused on the need to rapidly increase the protein supply for the population, in large part through the production of meat and other animal products. In line with Zlatarov, the congress—and the new communist regime—recognized the importance of studying food from a wide range of disciplinary angles. Science and technology, they noted, were critical tools for developing and implementing rational nutrition plans to amply fortify the socialist body.[89] Carefully edited and interpreted, Zlatarov's advocacy of higher amounts of protein in the Bulgarian diet was adopted, but minus its vegetarian foundation. This was directly in line with the Soviet party line under Stalin. The Soviet Union had moved toward a meat-based diet as the best way to provide protein and the "good life" to the masses. Any and all vegetarian underpinnings of food science—or a reliance on plant-based protein—would need to be jettisoned in the postwar push for protein parity and progress.

In the years from 1948 to 1990, Bulgaria's total raw meat production increased by 836 percent.[90] A good portion of this was for export, which reached unprecedented levels, but even for the average Bulgarian, the annual consumption of meat per person grew astronomically, from 3.5 kilograms in 1959 to 81.5 kilograms in 1989.[91] The bulk of the Bulgarian population went from eating meat only occasionally to consuming it almost daily by the late 1980s. Yes, one can talk about "meat shortages" and the lack of variety in meats that punctuated the period. Supply lines were certainly inefficient for meat, as for other supplies.[92] As elsewhere in the Bloc, the procurement of food supplies required local knowledge, connections, and ingenuity, to say nothing of hoarding and filching. On the whole, however, meat became widely available, both for home and for restaurant consumption. The new meat economy and diet represented a dramatic change for Bulgarians, and in a certain way the communists' aspirational meatopia was realized. This shift to a meat-heavy diet, as with bread, was achieved not just through new technologies and increased quantity of production, but also by a shift in form and substance. Prior to 1944, mutton and lamb were the most commonly consumed animal flesh in Bulgaria, with pork, beef, chicken, and goat trailing well behind. In the course of the communist period, however, pork became king. The mass production of pork became the most viable way to fulfill communist quotas as outlined in five-year plans, especially with the introduction of new breeds of larger, fattier pigs.

The path to pork plenty, however, was an arduous one. It would require postwar recovery and a far-reaching transformation in production and exchange. The mass butchering of animals during World War II had significantly reduced animal counts across the country. Higher rates of animal slaughter were tied to the exports of food to Nazi Germany, but also the need to eat animals when grain was scarce (such as during the 1942 and 1945 droughts). There was also the wartime need to plant grain to feed humans, rather than fodder to feed animals, so slaughter was necessary.[93] Wartime mobilization in a sense did not end, but rather transitioned into peacetime mobilization with the goal of postwar recovery and building socialism. The state played an even more central role in food production and trade, as it collectivized agriculture, including animal husbandry, but also took control of production and distribution chains. In 1947 small local butchers began to be nationalized, replaced by fewer, larger, centralized meat shops. A centralized administration for meat, Mesotsentral (literally, Meat Central), was set up in 1950 to coordinate animal collection and butchering stations across the country to supply local meat stores. An ethnic Turkish butcher from the Black Sea coastal city of Stalin (today's Varna) described this change in his correspondence with a relative in Istanbul that was passed on to Radio Free Liberty (and is now housed in the organization's archives). By 1951, he reported, some 160 small private butcher shops had been replaced by eighteen state-controlled shops, with all butchers then forced to join the local butchers' union, which had 244 members in 1951. Butchers were expected to adhere to state-imposed meat rations in these early years, which would ensure that everyone had equal access to meat, with a ration of (at least) two hundred grams a week. They were also allowed to sell any surplus meat after assigned rations were supplied, which made them the envy of other trades. Many butchers, he claimed, would take advantage of this privilege and swindle customers, altering their scales and distributing less than the minimum two hundred grams of meat a week or pawning off fat and gristle to fill rations. They could then sell surplus meat on the lucrative black market.[94] From the very beginning, the system combined a rigid structure of state-controlled distribution with limited incentives. This only served to spur gray- or black-market exchanges of meat (and other goods), which were in tension with and at times undermined official avenues of distribution. This would become a permanent (and problematic) feature of the socialist economy, not just for meat but for most goods.

In spite of such problems, there were also considerable successes in terms of achieving the state's objectives for meat production and consumption. From 1948 to 1958, raw meat procurement figures increased from 318 tons to 513 tons, with processed meats rising from 48 to 87.5 tons.[95] But that was only the beginning of efforts to ramp up the meat industry. In 1955, meat industry officials complained about the "decrepit" condition of existing meat plants, and laid out a fifteen-year plan for exponential growth, including new slaughterhouses with double the capacity and extensive refrigeration facilities.[96] This growth required that all aspects of the supply chain be improved, from advances in animal husbandry and inclusion of new breeds to the mechanization of processing and improvements in hygiene, storage, and transport. Even with the continual reports of challenges, problems, and deficiencies, the numbers were astounding.

Pigs played an important role in this momentous expansion of meat supply. Pork had become more commonly consumed in Bulgaria after 1878 as local wild pigs began to be domesticated in larger numbers and interbred with English Large Whites (Yorkshires) and German Edelschwein. In the 1960s, Bulgaria imported thousands of "pedigreed" Large White and Danish Landrace pigs from the USSR, Poland, and Sweden to interbreed them with local pigs, creating ever-fatter "races" of domesticated pigs.[97] These new super-pigs became an easy way to fulfill the plan in meat production, as they provided many more pounds of meat per pig. As a result, the numbers of sheep and cows rose slowly in this period, while the number of pigs rose exponentially.[98] In 1955, for example, officials projected a 125.4 percent increase in pig stocks, as compared to cows/buffalo, at 18.5 percent, and sheep/goats, at 23 percent.[99] Bulgarians still ate more mutton and lamb than the European average by the end of the period, but they had largely become pork eaters, not because of tradition but because mass production of pork was cheaper and easier—more so than for beef or mutton—and also because it allowed for overages to be exported to the USSR and, to a lesser extent, the Bloc, which were largely pork-oriented cultures.[100]

The collectivization of agriculture was an important piece of this puzzle both economically and culturally. Agricultural collectivization, which began as a "voluntary" process in 1947, became forced by the early 1950s (as outlined in chapter 1). These farms had an economic as well as a political and cultural function, as they served as conduits for political conformity, but also the eradication of "backwardness" in the countryside. This was directed toward all rural populations but had particular consequences for Muslim minority populations. On these new collective farms, Muslim minorities were expected to do their part in raising the new legions of pigs, despite their religious taboo against eating and breeding these animals. In spite of considerable resistance, local officials reported "successes" from rural Muslim districts, where "the attitude of laboring Turks toward pigs and pig farming has changed. Until recently they couldn't even speak of such a thing, and now it is part of their economic activities. . . . Even Turkish women are working on the pig farms."[101] In large part, the porkification of Bulgarian meat production accompanied the Bulgarianization of Muslim minorities, which peaked in the 1980s—precisely when pork production reached its highest levels.[102] As noted, however, Muslims were not the only, or even the primary, target of this turn to pork. The plan was to biochemically fortify the whole population, with meat to accompany, propel, and flavor progress. If the shift to mass pork production required that rural Muslims shed any former "superstitions" about pig breeding, it also required that Christians shed "superstitions" about the meatless fast.

Putting such plans into action required the enlistment of a veritable army of foot soldiers—food producers, cookbook writers, scientists, and various professional and home cooks—around the new cult of meat (and science). The shift to a meat-rich diet would require changes in food practices and, above all, the culture. A range of state-sponsored publications attacked the "irrational" and exploitative foundations of the traditional Bulgarian diet. Food science and technology departments rapidly developed across academic institutions, and in 1959 the Institut na khranene (Institute of Nutrition) was formed, with the goal of "rationalizing" food production and consumption.[103] The institute and its satellites employed a fleet of scientists, engineers, agronomists, and cooks, who studied everything from the biochemistry of foods to the mechanization of production and packaging techniques.[104] These facilities provided the research and development required to both justify and coordinate the plans to increase the production of meat and other foodstuffs throughout Bulgaria.

Tasho Tashev, a student of Asen Zlatarov, was the first director of the Institute of Nutrition, which carried out a ten-year study of the eating patterns of the Bulgarian population. As a follow-up to Zlatarov's and other interwar scholars' work, it set out to precisely track the caloric/nutritional "transition to socialism."[105] In such studies, the traditional Bulgarian diet was found to be severely lacking in protein, although it was improving year by year under the conditions of socialism. By 1962, an institute-supported study concluded that while Bulgarians had reached the optimal *caloric* intake, the "protein problem" still loomed large.[106] Various studies sponsored by the institute showed that even as Bulgaria's protein levels were reaching those in the "developed world," more than two-thirds of that protein was still derived from plant, not animal, sources; the stated goal was a 50/50 split.[107] As a result, nutritionists continually recommended a serious increase in the production and consumption of meat and other forms of animal protein.[108] The ultimate goal was 80 kilograms of meat consumed per person per year (roughly 0.2 kilos or 0.4 pounds a day), an objective that was realized only in 1988.[109]

The institute played a direct role in justifying and circulating a new meat-heavy food science in the service of this goal.[110] It produced or provided fodder for a range of works on diet and cuisine that decried religion-based dietary restrictions or cycles of feast and fast as outmoded, "superstitious," and "irrational." In a general work on Bulgarian nutrition, for example, Tashev noted with exasperation that the Bulgarian peasant diet had its longest meatless fast in the spring, right before Lent—precisely when energy was needed for the spring planting and other activities. They then ate more meat (and hence fats and proteins) than they needed in the winter, which was the down season for rural labor.[111] These assessments were clearly geared toward changing the eating patterns of rural populations, but such populations were also entering the city in droves. They were the new urbanites in a period of mass urbanization, who needed to be molded into meat-fortified socialist men and women. For this to happen, meat could no longer be the stuff of "feasts" or special occasions; it had to be a regular part of the everyday diet.

To this end, the shift toward meat was embedded in a new socialist culinary culture infused with science. By prescribing a gram of this and a pinch of that, cookbooks could precisely prescribe "rational" eating, in and beyond meat. But before one even got to the recipes, most cookbooks of the communist era commenced with sections on the science of food, in which the concept of protein and the advocacy of meat were central. The earliest socialist-era cookbooks, in particular, dedicated an inordinate amount of space to the science of food as part of their mission to increase people's intake of calories and protein, and hence their productivity and wellness. One such cookbook, the 1953 *Rŭkovodstvo po obshtestveno khranene* (Management of public nutrition), was coauthored by a professional cook and a food scientist as a comprehensive guide to the science and culinary techniques of food. Significantly, the first 100 of the book's 250 pages are a kind of treatise on the importance of proper nutrition, which is repeatedly cited as necessary for human health, energy, and labor.[112] *Rŭkovodstvo*, like most of the cookbooks that followed, detailed the fundamental biochemical elements of the productive body, with protein specified as "the primary building block in our food" and essential for "mental and physical labor."[113]

FIGURE 2.2.   Bulgarian shoppers line up while waiting for a meat shop to open (circa 1970s, location unknown). Photo by Chris Niedenthal//Time Life Pictures/via Getty Images

The scales for protein were also tipped by the weight of ideological and historical legitimacy and consequence. The words of Friedrich Engels from his 1878 work *Anti-Dühring* provided ideological ballast to the protein project. In this survey of the state of science in his time, Engels noted that "protein was life," and hence inadequate protein was tantamount to "death." Quotes from his 1883 *Dialectics of Nature* drove the point even further: "With due respect to vegetarians, man did not come into existence without a meat diet," and "meat has played a major role in the historical development of man, because it contains in almost ready form the most important elements needed by the organism."[114] According to the authors' reading of Engels, protein (and especially meat) could be seen as a kind of cornerstone—the material fuel—for historical progress. With such justifications in place, they advanced the notion that "animal-based foods are the most important in our cuisine," before outlining preparation methods for all manner of meat in splendid detail.[115] In the pages that follow, meat-based recipes and menus abound. There is no lack of vegetable dishes, to be sure, but they are not referred to as "vegetarian" or "*postni*" (meat-fast or vegan), as they were in interwar cookbooks. In this period, vegetarianism as such was off the table.

The audience for *Rŭkovodstvo* was in large part professional cooks, but clearly the book was also targeting home cooks, primarily women. Cookbooks and columns in women's magazines that printed recipes were an excellent vehicle for spreading the gospel of meat. From the earliest postwar years, women had a special role in the implementation of "rational" eating, not just their own, but that of their family. They were expected to purchase, preserve, prepare, serve, and manage food in a whole array of new ways. For most peasant women coming from rural environments—the bulk of Bulgarian women—this was an entirely new way of cooking and eating, based on knowing and calculating nutritional elements. In many respects, socialism had ushered in a new era for women across the region, with increased educational and professional opportunities. Women were expected to work outside the home, and mechanized food production and distribution, along with "communal eateries" of various forms, were supposed to "liberate" them from the labor of the field and kitchen. In practice, however, women still had the primary responsibility for the bulk of the work at home, and cookbooks were written with this in mind.[116] For men, cooking was either a profession (for chefs) or a hobby. In contrast, it was women's responsibility to implement the socialist plan for new modes of eating.

But expectations for women went well beyond the rational and nutritional. They were increasingly responsible for the creation of modern and "civilized" domestic utopias. Socialist-era cookbooks clearly had an educational and even civilizing mission, but they were also part of promoting the socialist good life in the 1960s and 1970s. This was a period of novel and growing abundance, but also, after Stalin's death in 1953, of greater attention to consumer desires.[117] Interestingly, the second edition of the cookbook, which came out in 1955, bore a much folksier title: *Nasha kukhnia* (Our cuisine). This was perhaps an indicator of a subtle shift in orientation, a softening of the ideological front in the wake of Stalin's death in 1953.[118] In this new version, the section on "Public Nutrition" was greatly abridged, and the quotes from Engels were edited out. Many of the same ideas and principles are present, however, with 200 grams of meat a day in the suggested daily menu.[119] This is a stark contrast to the postwar ration of 200 grams *a week*, but it still adds up to a mere 10.4 kilograms a year, a far cry from the later goal of 80 kilograms a year. In later cookbooks, such as the 1956 *Sŭvremenna*

FIGURE 2.3. Advertisement (from a calendar) for the meat-processing factory Rodopa in Blagoevgrad, Bulgaria, in 1981. http://www.lostbulgaria.com/?p=4283.

kukhnia: 3000 retsepta (Modern cuisine: 3,000 recipes), nutritional information was pared down even more to make room for the thousands of recipes
that women were somehow supposed to have time to prepare.[120]

Communists continued to advocate a regime of "rational" consumption
relative to bourgeois excess or "irrational" peasant feasting and fasting. But
they were not against pleasures of the flesh per se. In fact, such pleasures
in the correct amount were seen as a deserved reward for citizens' labor.
Meat was assumed to be part of the new rational diet, but it was also part
of a larger vision of the socialist good life that provided a tasty slice of the
traditional feast, ideally every day. In part this was about home preparation
and consumption, as cities and towns had a growing number of meat shops
whose shelves were laden with new kinds of processed meats, including
sausages and salamis. At the same time, the development of an affordable
restaurant culture contributed markedly to the shift toward regular meat
consumption. The state tourist industry, Balkanturist, coordinated the establishment of new restaurants across Bulgaria, at which meat was centrally featured. The new prevalence and affordability of the restaurant were part and
parcel of making meat both available and easy to consume—no butchering,
plucking, or deboning required. Meat dishes were a mainstay on the menu at
all restaurants and street stalls, which mushroomed across Bulgaria in numbers far exceeding those in other socialist countries. By 1968, 40 percent of
food in Bulgaria was consumed in restaurants, which outstripped all other
countries in the Bloc, including Hungary (33 percent), Romania (26 percent),
and the USSR (18.9 percent).[121] In part this was because of Bulgaria's status
as a tourist destination. Admittedly, the majority of visitors continued to
be from the Bloc, for whom Bulgaria was significantly more affordable and
accessible, but the number of Western visitors was also on the rise. Bulgaria
had miles and miles of southerly sandy beach destinations, where the state
built veritable cities of resort hotels. With Balkanturist in charge of this new
tourist empire, the number of international tourists rose to one million in
1965, and to three million by 1972.[122] All those foreign bodies needed to be
fed, and according to Bulgaria's own statistics, Westerners ate meat regularly
and would expect to maintain this diet on Bulgarian soil. In the Burgas district on the southern Black Sea coast, a number of large meat plants were
built and regularly revamped, with the goal of producing the needed meat
for tourists.[123]

The offerings at such restaurants varied widely, from very simple to more
luxurious fare, with meat as a mainstay. As a British traveler marveled in his
1966 travelogue, "There are endless varieties of pork and veal, *schnitzel* . . .
brains, kidneys, liver, lamb and chicken." From chicken Kiev to roast suckling

pig or "Bulgarian" dishes like "various kebabs of minced meats and herbs," there was no lack of meat on the menu.[124] In the context of the restaurant and cookbook, it is worth noting that all manner of animal organs were routinely found, including grilled brain, liver soup, and the popular *shkembe chorba* (tripe soup). While such variety was not always present, there were four meat dishes that could be found on virtually all menus—*kebabche, kiufte,* shish kebab, and pork cutlet. The *kebabche*, a grilled cylinder of ground pork, and the *kiufte*, a round patty of ground meat, were available not only in restaurants, but also from street vendors and at smaller *skara* (grill) stands.[125] *Kebabche* and *kiufte* were not just delicious—they were well suited to the communist period's penchant for pork, especially as one could use various cuts and scraps to make them (sometimes mixed with lamb or beef). If pork had been slow to enter Bulgarian cuisine before 1878, by this time it became king through its integration into these and other dishes, a mainstay of restaurant menus. This dining experience—awash in meat—went beyond nutritive fortification. It was also part of the new modes of leisure, which were increasingly oriented toward pleasurable consuming experiences.

Finally, meat was connected to the amplified articulations of national identity that characterized the late socialist quest to hang on to legitimacy. The first postwar cookbook to openly call itself "national" was *Bŭlgarska natsionalna kukhnia* (Bulgarian national cuisine). Published in 1978, it was chock-full of meat dishes. It featured more than eight different varieties of *kiufte*, made with pork or veal, but also onion, egg, bread crumbs, red and black pepper, and salt.[126] The obligatory detailed treatise on nutrition was compressed into an extremely short survey on the positive changes in Bulgarian eating habits since the advent of communism. The pre-communist diet had been "poor," it noted, characterized by a lack of meat and too much bread. In contrast, the communist-era diet had met 100–103 percent of protein needs, with a major boost in animal protein consumption.[127] Quickly moving beyond nutrition to gastronomy, *Bŭlgarska natsionalna kukhnia* offered an elaborate history of "Bulgarian cuisine," from picking through the archaeological finds of the earliest inhabitants of the region to foreign travelogues and Bulgarian National Revival writings. Meat played an important role in this constructed "Bulgarian" culinary heritage and pedigree, which traced local animal husbandry back to the ancient Thracians. Meat was also cited as the central food in the diet of the nomadic Proto-Bulgarians, who established themselves in the region in the seventh century.[128] While confirming the Bulgarian "tradition" of copious meat eating, the book also included plenty of recipes for salads and other meatless dishes. Salads were clearly of more recent invention (for more on this, see chapter 4). But the book also listed a range of old and

FIGURE 2.4. Interior view of a cooperative pig farm in Elkhovo, Bulgaria, circa 1960s. Keystone-France, via Getty Images.

new plant-based dishes, many marked as *"postni"*—a nod to the Orthodox tradition of the meatless fast.[129] This was a subtle yet significant shift, part of an undercurrent of change in the 1970s that had important implications in terms of the meaning and place of meat in the Bulgarian bioimaginary.

The Vegetarian Swerve

Meat provided an important staple for a late socialist culture of leisure and pleasure that was meant to exemplify the coming of a utopian future. In the last decades of socialism, it was central to state efforts to use consumer offerings, along with nationalism, to build consensus and legitimacy across the Bloc. And yet, for critics within the system, such consumer pleasures could be seen as a threat to, or *replacement* for, such a future. In the West, critiques of rampant consumerism surged in the 1960s and 1970s, as the new generation rebelled against the materialist values of their parents. Among other things, this movement was grounded in a critique of the "meat-laden, macho 1940s and 1950s" and was intertwined with new (or revived historical) food practices, including vegetarianism.[130] Under late socialism, change was also afoot, both from within the communist establishment and among countercultures of various sorts. But few looked to vegetarianism as part of their critique of the older generation, or the consumerist direction of late socialism.

An important and often overlooked example of a vegetarian turn in the Eastern Bloc came from within the highest levels of power in Bulgaria, namely the country's enigmatic "first lady" in the 1970s, Liudmila Zhivkova. The daughter of Todor Zhivkov, Zhivkova remains an enormously contro-versial figure—more like a phenomenon—who made a distinctive mark on Bulgarian state socialism in that decade. After her mother's untimely death in 1971, Zhivkova rose to prominence with the help and approval of her father, who held his well-educated daughter in high esteem—she had received a PhD at Oxford.[131] He appointed her as head of the National Committee of Art and Culture in 1972 and to the Politburo in 1979. In both positions she had a tremendous amount of power and resources at her disposal, and became known for her role in promoting Bulgarian culture both at home and abroad, while cultivating global cultural-*cum*-diplomatic ties.[132]

Zhivkova's power raised many eyebrows in the higher echelons of com-munist officials in both Bulgaria and the USSR, in large part because of her "alternative" ways of thinking and living. Beginning in the early 1970s, and especially after a bad car accident in 1973, Zhivkova had made a decisive turn toward a range of esoteric spiritual schools of thought and practices. One of her main influences was Agni Yoga, a yogic system of exercise and thought developed and practiced by the famous Russian couple Nikolai and Elena Roerich. The Roerichs had become expatriates from the Soviet Union in the 1920s and from then on lived in a variety of places, including Finland, the United States, the United Kingdom, and India. They were influenced by the theosophical movement, cofounded in 1875 in New York by another Russian émigré, Helena Blavatsky. In 1921 the couple founded the Agni Yoga move-ment, calling for strict vegetarianism and abstinence from alcohol, as well as frequent fasting. The Roerichs' intermittent presence in the Soviet Union was tolerated in the freer environment of Leninist Russia, but once Stalin had established power, they were denounced as "sectarian." Nikolai (who died in

India in 1947) was slowly rehabilitated in the post-Stalinist period, but even then the Soviets recognized him as an artist (he was a prolific painter) and not as a spiritual thinker. Thus it was quite bold for Zhivkova to openly embrace Agni Yoga and proclaim 1978 the year of Nikolai Roerich, in spite of disapproval from the Soviet Union and a number of Bulgarian communists.[133] She organized a series of events that year, including exhibits and symposia to which she invited from India the Roerichs' son Svetoslav and his wife, Devika Rani (the actress and granddaughter of Rabindranath Tagore).[134] Zhivkova nurtured Bulgaria's ties with India in this decade, maintaining close contact with Indira Gandhi. She made frequent visits to India, where she had (ultimately unfulfilled) plans to financially support the building of an institute in the Roerichs' name.[135]

Zhivkova's spiritual (and vegetarian) inspiration, however, was tied to deeply local sources. She promoted Bulgarian culture and history in all forms, in Bulgaria as well as abroad. Her efforts in this regard were unprecedented within the Bloc, almost scandalously so. The funds she spent in order to take exhibits, folk troupes, and other examples of Bulgarian culture around the globe were astronomical. And yet, she personally tapped into schools of thought from the region's past that were associated with bodily restraint. Indeed, Zhivkova was a driving force behind academic and popular interest in the ancient Thracians, and especially the restrained vegetarian cult of Orpheus.[136] She also looked for inspiration to the eleventh-century vegetarian Bogomils and the early twentieth-century Bulgarian White Brotherhood. Zhivkova often wore the telltale white of the Dŭnovtsi, along with a controversial head covering that resembled a turban. Far from alone in her spiritual pursuits, she had a kind of "salon" of like-minded Bulgarian intellectuals, and propagated the idea that Bulgaria had a special spiritual mission, to create a cultural synthesis of East and West with global significance.[137] Zhivkova was keen on launching such a mission through various initiatives that transcended her personal life. With her own bodily practices as her foundation, she infused her public speeches and publications with idealized notions of beauty and harmony. Her rhetoric transcended religion but also pushed the boundaries of Marxism in a way that was puzzling and even shocking to many of the party bureaucrats—not to mention Bulgarian audiences—in her orbit.[138] And yet she integrated a Marxist vision of the future into her speeches and even her private musings; communism, at least in theory, was consistent with her utopian rhetoric.

Zhivkova by no means made vegetarianism an official policy in socialist Bulgaria; nor did she openly advocate it in her speeches or publications. But she also made no secret of her way of eating and living. At least a handful of

top communists reportedly went meatless to impress her, if only in her presence.[139] More importantly, in stark contrast to her father—a well-known big-
game hunter—Zhivkova created an alternative narrative and model for late
socialist progress that opposed rather than valorized meat consumption.
Needless to say, she did not approve of her father's hunting, though that
did not seem to come between them.[140] Nor did she embrace that aspect of
the Bulgarian late socialist good life that revolved around a heaping plate
of *kebabche* and a beer, or roasted lamb and a glass of wine. For Zhivkova,
personal purity through vegetarianism and abstinence from alcohol was a
building block of collective fulfillment, which in the end could bring the
arrival of ripe communism.[141] In a sense, her Marxist model tapped into
socialist thought that denounced predation and pleasure-seeking in favor
of "harmony" and "beauty." It also looked back to the turn of the century
and interwar critiques of the West that were entangled with a renunciation
of meat.

To be clear, Zhivkova's realm of influence was primarily cultural policy.
Meat production and consumption did not slow in the slightest during her
years of influence; on the contrary, it continued to rise. And yet Zhivkova
was enormously influential in the flowering of a kind of alternative culture
and science. Research, publications, and exhibitions on the Bogomils and
Thracians—and Nikolai Roerich—were one result of her sway at the highest levels. But she also intervened on behalf of the remaining Bulgarian followers of the White Brotherhood, as well as a subculture interested in yoga
and other "esoteric" practices.[142] She was responsible for the publication
of works by a number of people who had been regarded as too "alternative" for publication in the past. Most apropos to food were the complete
works of Petŭr Dimkov, a prominent guru of Bulgarian folk medicine who
was influenced by the ideas of Tolstoy and the White Brotherhood.[143] His
three-volume set of books published in 1977–79 included detailed sections
on "food as medicine," in which he openly advocated vegetarianism and
fasting.[144] In 1980 Alexander Belorechki published a book explaining and
advocating vegetarianism (albeit for those with illnesses), and a range of
other cookbooks began to include more dishes labeled as *"postni"* (associated with the Orthodox vegan fast).[145] In the end, Zhivkova's alternative
vision struck a chord with many who were inspired by her aesthetic and
antimaterialist vision. At the same time, she posed a threat to many within
the system, whose legitimacy was on extremely thin ice by the beginning
of the 1980s.

In 1981, Zhivkova was found dead in the Zhivkovs' villa outside Sofia.
She was only thirty-eight years old. The official cause was an aneurysm,

but there were a number of strange circumstances surrounding her death, which is shrouded in mystery to this day. Many believe that she was a victim of the KGB or internal communist operatives who viewed her as too great a liability to a system that was struggling to stay relevant and viable.[146] Some speculated that her vegetarianism and fasting had killed her—that she had starved herself to death—or at least that her meager eating had contributed to the failure of her "frail constitution."[147] Zhivkova was sincerely mourned with the necessary pageantry, but many of her initiatives were subsequently reversed or scaled back. Still, her legacy could not be totally pushed aside.

In 1981, for example, a more inclusive set of Asen Zlatarov's lectures was published as a kind of textbook for students of Bulgarian food science. This included his essays on vegetarianism, decrying the "aggressive nature" of meat eaters, and especially the "predatory nature of the West" as the society that ate by far the most meat per capita in the 1920s (when he wrote these lectures).[148] The irony of this was that by 1981, when the lectures were republished, much of the socialist world was consuming as much or more meat per capita than the United States. If anything, meat consumption was on the decline in the United States, because of widespread concerns about the health effects of cholesterol, concerns that were still prominent in the Eastern Bloc. Even as Zlatarov's works gave soundly leftist reasons for shirking meat, the production and consumption of meat rose unabated in Bulgaria until it finally reached the state's goal of eighty kilograms per person per year. In spite of the Zhivkova effect, the 1980s was the most carnivorous decade in Bulgarian history.

In 1989 communism collapsed across Eastern Europe, in a domino effect that began in Poland and Hungary and ultimately swept the Bulgarian communists out of power. The transition to capitalism and democracy has been a painful one for large swaths of the Bulgarian population. Interestingly, that "pain" has been clearly palpable in animal husbandry and the meat industry, which entered into a free fall after 1989: the production of meat decreased by 65 percent from 1990 to 1995. While the economy, and animal husbandry, stabilized somewhat in the late 1990s, this industry has never recovered. Bulgaria is no longer the mass meat-producing (or eating) country of the past, at least not on the scale it once was. In fact, in the years since Bulgaria joined the European Union in 2007, a large number of its meat-processing facilities have been shut down as "unhygienic."[149] Consumption has also fallen precipitously: by 2000 it was about half the 1989 level.[150] For many, this is part of the painful new reality of market conditions and the lack of subsidized food. But perhaps less meat is also a good thing, a turn toward a more economically and environmentally sustainable future. In the post-socialist period, new meatless paths have emerged—including the Orthodox meatless fast and a new wave of followers of the White Brotherhood. If the communist transformation has left its mark on a meat-eating population, meatless traditions also have had considerable staying power in Bulgarian history.

Bulgaria has long been a center of animal husbandry and a regional supplier of meat to its various allies. Animals, and meat, were a key part of the local economy and culture, but the majority of Bulgarians ate relatively meager amounts of meat until the communist era, when meat went from an occasionally consumed food to part of the socialist everyday. In that period, Bulgaria—like the rest of the Eastern Bloc—was engaged in a meat race that both reflected and drove the postwar transformation of society as a whole, tied to industrialization, collectivization, and other changes in agriculture and animal husbandry. This followed a global pattern in many respects, but in others it followed a uniquely socialist and Bulgarian playbook. For the Bulgarian Communist Party, and for the Bloc as a whole, increased meat production and consumption were essential to the long-term political goals of "building socialism" and eventually achieving a communist utopian future. Protein was needed to fuel workers' and peasants' bodies—to power the musculature that would swing the proverbial hammer and sickle. If workers needed meat as physical fuel, it also could reward them for a hard day's work and provide ballast for the socialist good life in the here and now. Meat was central to the aim of catching up to the West, a process that was tied to nutritional targets that fueled the Bulgarian and Bloc (bio)imagination. This turn toward meat in the Bulgarian diet was not entirely new, but the pace was unprecedented. And as a result, the postwar shift to a more meat-based diet demanded economic and *cultural* transformation, such as the privileging of meat eating in cookbooks and restaurants, and the repression of vegetarian writings and practices that had taken hold in the interwar period.

Such vegetarian impulses saw the light in the 1970s but could not overcome the powerful forces that were intent on building a meat-eating society. Those forces were largely successful, but at what cost? For as meat production, consumption, and export numbers soared, grain production for human consumption was inadequate. In spite of efforts by the state and the Bloc to reduce consumption of bread and grain, demand did not abate. Instead, meat supplemented, rather than supplanted, the local bread diet. If anything, it over-fulfilled dietary needs, thereby creating an unsustainable food system that Bulgaria could ill afford. In essence, the socialist cult of meat played an important role in the unraveling of the system, in both its abundance and, at times, its shortage. Capitalism, however, was not the clear answer. Indeed, post-socialist nostalgia was grounded, among other things, in memories of a time when meat and other products were plentiful and cheap, and when the local meat industry was strong, given its captive market. At the same time, vegetarian (or periodic fasting) traditions in Bulgaria have reemerged. In the end, these meatless (or meat-light) traditions might offer Bulgaria, and perhaps the world, the most sustainable and equitable future.

 Chapter 3

Sour Milk

Long Life, the Future, and the Gut

The 1970s Bulgarian how-to for the home *Kniga za vseki den i vseki dom* (The book for every day and every home) poses the enduring question, "To be *dŭlgoletni* [long-living], isn't that every person's dream?" After a polite nod to socialist progress and medical advances, this popular late socialist encyclopedia of the everyday turns quickly to food for a detailed answer. History and science scaffold the discussion, which looks to scholars, from Plutarch to Pavlov, who explored the possibilities of extending life through "regimes of eating."[1] Russian scientist Ivan Pavlov was a particularly important sage in socialist-era food writings, given his famous experiments on dogs, bells, and digestive responses. But there was a second, and perhaps more critical, prophet whose name was constantly evoked in popular and scientific works on food. That was Nobel laureate Ilya Mechnikov (often transcribed as Metchnikoff), whose turn-of-the-century scientific hypotheses and personal health practices looked to Bulgarian *kiselo mliako* (yogurt—literally "sour milk") as a curative fix for the intestinal ills—the rotting gut—of the modern age. Mechnikov's widely publicized belief in "Bulgarian sour milk" as the biochemical elixir of long life was foundational to the scientific-myth complex that became linked to yogurt and has had a dynamic afterlife both in Bulgaria and globally.

What exactly did Mechnikov mean by "Bulgarian" sour milk? Yogurt, after all, is not unique to Bulgaria. Rather, fermented milk has deep historical roots in the Eurasian yogurt belt—from the Balkans, via the Caucasus and Central Asia, to the Indian subcontinent—or what historian Anne Mendelson calls "Yogurtistan."[2] While aware of this range and reach of milk fermentation practices, Mechnikov nevertheless believed that the particular microorganisms found in Bulgaria, the bacterial species *Bacillus bulgaricus*, had the "most useful" properties for replenishing colonic flora and promoting longevity.[3] In a period before a Bulgarian national cuisine had been articulated, Mechnikov deemed this broadly regional substance as "national"—based not on its flavor, but on its distinct biochemical makeup. Mechnikov's advocacy spurred an early twentieth-century international food fad based on "Bulgarian sour milk" or yogurt. While the fad gradually fizzled out, yogurt (in various forms), albeit without the "Bulgarian" label, became an enduring part of the Western, and global, diet.

In Bulgaria, the history of yogurt is equal parts tradition and transformation. But it was only under socialism that yogurt became a critical ingredient in the Bulgarian diet and bioimaginary, part of the recipe for a communist future. Recognized not only for its probiotic properties, but also as an important source of protein (along with vitamins and minerals), yogurt underwent a massive increase in production and consumption in the course of this period. State-sponsored scientists, cookbook writers, and public health

FIGURE 3.1. Russian microbiologist Ilya Mechnikov at his microscope, 1913. Bibliothèque nationale de France, gallica.bnf.fr.

and home economy specialists extolled the flavor and the nutritional, diges-tive, and even historical-cultural properties of this milky wonder. Bulgar-ian yogurt (until recently) was never presweetened, and its natural sour and gamy *terroir* allowed for its flexible use in sweet and savory dishes, drinks, and baked goods. Prominently featured in recipes and recommended diets for "common health problems," including high blood pressure and heart dis-ease, yogurt helped bolster the enduring belief in Bulgaria that food was not just for sustenance; it was medicine.[4] As noted in the 1973 edition of *Kniga za vseki den*, for example, "Mechnikov believed that regular consump-tion of yogurt, fermented with *Bacillus bulgaricus*, furthered longevity. Years of observation have shown this idea of the great scholar to be correct, and therefore we recommend the daily consumption of yogurt."[5] Here, as in a spate of socialist-era writings on the wondrous properties of dairy products in general and yogurt in particular, the stated goal was optimal health, pro-ductivity, and long life.

Significantly, in communist-era writings on yogurt, history was as impor-tant as biochemistry in securing yogurt's traditional form and place in the everyday. In a range of sources, narratives abound on the ancient origins of Bulgarian yogurt, its delicate bacterial flora and flavor. Scientists, historians, and cooks cast yogurt as a national treasure, to be widely consumed but also preserved in its *traditional* form. If Mechnikov catalyzed a new mythol-ogy around the scientific properties of yogurt, Bulgarians also looked to the ancient origins of this microbial fountain of youth to solidify its newly prominent place in the modern diet.

The Sour Milk Sensation

A 1912 feature story in the *New York Times* titled "Metchnikoff Confirmed in His Theory of Long Life" starts with the following:

> In a little village in Bulgaria there is living to-day Baba Vasilka, who has reached the goodly age of 126 years and is declared to be the "old-est woman in the world." She has a son, Tudor, a youth of 101 years, active and vigorous, and altogether likely to add several more decades to his life. . . . Were Baba and Tudor Vasilka living in any other country than Bulgaria, they would be objects of curiosity and solicitude. . . . In Bulgaria, however, the age reached by Baba Vasilka has not been uncommon.[6]

Under a blurry black-and-white photograph of Baba—which simply means "grandmother" in Bulgarian—and her son, there are several insets depicting petri dishes teeming with long, thin *Bacillus bulgaricus*. As the article explains, Mechnikov's theory of longevity had attracted the interest of a number of scientists, who gathered evidence on Bulgarians and other "races that live to a hundred and more."[7] These scientists had now "confirmed" the theory that the lactic acid found in Bulgarian "sour milk" helped to counter the ill effects of what Mechnikov called "putrefaction" of the gut—the toxic rotting of foods trapped in the large intestine. The 1912 article was one of many that confirmed Mechnikov's Bulgarian yogurt thesis, which caused a huge stir in the scientific and popular health community in the years following his 1908 Nobel Prize in Physiology or Medicine.

To be clear, Mechnikov's Nobel Prize had nothing to do with fermented milk. Rather, it was awarded for his research in immunology, and specifically his discovery of phagocytes—cells that ingest (and destroy) foreign particles in the body during an infection. A zoologist who worked across various fields, Mechnikov pursued an expansive array of scientific interests. Originally based at the University of Odessa, he was invited by the famous French microbiologist Louis Pasteur to relocate to Paris in the late 1880s. Pasteur's own early research was instrumental in the science of fermentation, namely documentation of the fact that living microorganisms in milk and other foods made them decompose, turn to alcohol, or go bad.[8] But far from singing the praises of fermented food, Pasteur was credited with being at the forefront of the modern war on germs. Indeed, he and his institute played a critical role in proving germ theory, which had a revolutionary effect on medicine and food science, with particular significance for milk. Pasteur is most famous for developing the process of "pasteurization," the heating of foods to kill harmful pathogens. His discoveries propelled the increased consumption of "fresh" milk in its "pure" pasteurized form in the Anglo-American diet by the turn of the century.[9] Mechnikov's enthusiasm for bacteria within "sour milk," then, ran counter to developments on the science of germs, and especially their potential harm in milk, that had been established in the West.

In contrast, Mechnikov looked to sour milk's bacteria as a potent tool for bringing ecological balance to the body via the gut, an altogether new idea at the time. As he himself aged within the hallowed walls of the institute famous for milk science, it was perhaps natural that his research interests turned to aging and milk. By the time he rose to fame with the announcement that he had been awarded the Nobel Prize, Mechnikov was firmly preoccupied with aging, senility, death, and yogurt. As he argued in his last book, *The Prolongation of Life* (1907), the process of bodily degeneration was a preventable disease caused by noxious poisons festering in the gut.[10]

He hypothesized that in a plethora of diseases, infirmity, senility, and early death, bodily degeneration could be arrested by introducing "natural enemies" and creating a balance in the "flora" in the large intestine. Mechnikov's conclusion came from a long career spent studying animals, more specifically his observation that animals that lacked large intestines, such as birds, lived much longer than larger mammals with intestines. After outlining the theory in detail, he proposed a solution, based on his own observations and ongoing studies at the Pasteur Institute. Among his key observations was that there were an inordinate number of centenarians in Bulgaria, and the regional yogurt belt more widely, where some form of fermented milk was regularly consumed. Yogurt, he concluded, was a kind of fountain of youth because of its biochemical properties, and he consumed it every day thereafter until his death in 1916.[11]

Mechnikov's theory of the digestive holy grail, however, became scientifically grounded only when the curative bacterium in fermented milk was discovered by one of his contemporaries. In 1905 Stamen Grigorov, a Bulgarian scientist and physician, was the first to view what came to be called *Bacillus bulgaricus* through a microscope while studying medicine under the famous microbiologist Dr. Léon Massol (1838–1909) at the University of Geneva. According to later sources, Grigorov famously hauled a number of traditional ceramic vessels, most suited for making and storing the "sour milk," to Geneva from Bulgaria. One can only imagine his gleeful reaction upon seeing the yogurt's rod-shaped bacteria casually swimming across his microscope slide, for this was indeed an important discovery. At his adviser's urging, Grigorov presented and published his findings in Paris at the Pasteur Institute that same year. At some point during the visit, he bestowed a clay vessel full of Bulgarian yogurt upon his most famous, and vitally interested, talk attendee—Ilya Mechnikov himself.[12] In honor of Grigorov, this particular variety or family of lactic acid was named *Bacillus bulgaricus* (or *Lactobacillus bulgaricus*); in fact, the name is often followed by "Grigorov" in early scientific references. Mechnikov, duly impressed by the discovery, offered Grigorov directorship of the Pasteur Institute in São Paulo, Brazil.[13] But Grigorov chose to return to Bulgaria to continue his research, living out his life in relative obscurity as a country doctor. In contrast, Mechnikov became almost a household word—a sensation in the international press—along with Bulgarian sour milk.

Not long after he was awarded the Nobel Prize, Mechnikov's name lit up the popular and scientific press in the United States, the UK, and well beyond. But both kinds of sources seemed to veer away from his discovery of the important but somewhat arcane phagocytes and toward his claims about aging, the gut, and his daily diet of Bulgarian yogurt. A wide range of publications, including, as noted above, the *New York Times*, sang the praises of Mechnikov's "confirmed" discovery. In the *Independent*, Edwin Slosson—an American chemist, journalist, and popular science writer—called Mechnikov one of the "Twelve Major Prophets of Today," "the worthy successor of Pasteur."[14] "Sour milk" science was also greeted with interest and support by a wide range of popular and scientific publications, including *Scientific American* and the *American Journal of Medical Sciences*.[15] As a 1910 article in the London *Times* noted, "The present vogue of sour milk for dietetic and medicinal purposes is an example of the vindication of popular customs by science."[16]

But Mechnikov's notoriety and commitment to "sour milk" also brought skeptics, naysayers, and charlatans out of the woodwork to scrutinize—or profit from—the new wonder substance. For some there was revulsion at the thought of consuming "sour milk" in the increasingly hygiene-obsessed Anglo-American world. Serious concerns about pathogens in milk, and particularly their connection to infant mortality, gave rise to a widespread movement to sterilize, pasteurize, and refrigerate milk.[17] Even traditional fermented forms of milk, such as buttermilk, began to fall out of favor, as freshness and "purity" became the order of the day.[18] A 1910 article in the American publication *Health* decried the "Sour Milk Fetish," claiming that the "use of sour milk is wrong in principle and in fact." The author denounced those who rushed to profit from these "deadly microbes" when milk was "pure and safe" only in its fresh form. Ending with a sarcastic flourish, the article's anonymous author, under the pen name "Herald of Health," concluded, "I should go to Bulgaria and investigate the fairy-tale of the great longevity acquired by the inhabitants through their liking for milk-vinegar."[19]

For better or for worse, Bulgarians—and other peoples of the "East"—captured the Western public imagination amid this tangle of claims. A 1906 article in the *Phrenological Journal* by a Dr. C. H. Shepard, for example, endorsed the "prescriptions" of Mechnikov, which promised to "exterminate the deadly army of microbes which has installed itself in the citadel of our being." Echoing Mechnikov, Dr. Shepard looked to "Bulgaria, where the inhabitants are endowed with remarkable virility," as a result of "Bulgarian ferment."[20] He lauded "the customs of primitive nations," like the "Orientals" of the Ottoman lands, which were "more scientific than our own." The Bulgarians, it should be noted here, were still technically under Ottoman rule until 1908, albeit within an autonomous principality. Perhaps their liminal status as "Oriental" and yet white and European appealed to many in a period when "purity" was often connected to racial categories. At the same time, adulation of the gustatory wisdom of this distant "Oriental" peasant nation was at times combined with a backhanded dig at the Balkan region, which was increasingly associated with sectarian violence. In a 1910 article in the *Youth's Companion* on "Uses of Sour Milk," for example, the author notes that "The longest-living people in Europe, provided they escape death by knife or bullet, are the Slavic races of the Balkan Peninsula, who live very largely on bread and sour milk."[21]

As a result of the "viral" Mechnikov moment, yogurt made inroads into early twentieth-century American and Western European diets. It was all the rage, for example, among the purveyors of alternative notions of health like Dr. John Harvey Kellogg, the well-known American vegetarian and proponent of whole grain foods. Kellogg advocated the regular consumption of "Bulgarian yogurt"—and even promoted its use in enemas—to cleanse the colon of the "putrefaction" presumably caused by the consumption of animal flesh.[22] He was credited with saying, "Balance your intestinal flora, and you'll live as long as the rugged mountain men of Bulgaria!"[23] Yogurt consumption, "Bulgarian" or not, also began to spread to Western Europe. The French were particularly open to the notion of yogurt taking its place alongside modern "fresh" milk, perhaps because France was home to the Pasteur Institute. The other factor was the Danone Company, founded by Isaac Carasso, a Sephardic Jewish businessman from the city of Thessaloniki, which was transferred from Ottoman to Greek rule only in 1912. From this city in Bulgaria's backyard, the Carasso family moved to Barcelona and began a yogurt business, having purchased some "Bulgarian yogurt starter" at the Pasteur Institute in 1919.[24] The Danone Company, which moved to Paris in 1923, was born out of the marketability of the marriage of "tradition" (in this case foreign or "exotic" tradition) with science, which came to be a common—if intermittent—feature of patterns of food science and marketing (or advocacy) in the West. This trend took on its own forms and rhythms in the Bulgarian context.

Oriental Sour Milk

The Balkan region was at the westernmost tip of "Yogurtistan," the aforementioned ancient fermented-milk belt that extended from the Balkans to the Indian subcontinent. Most historians of yogurt agree that the regular consumption of fermented milk products originated in Mesopotamia and Central Asia and spread outward from there with migrations and conquering armies. As far back as 8000 BCE, peoples of the region discovered that the milk they were harvesting from local animals—sheep, goats, cows, camels, horses—curdled in the summer heat. The fermentation process allowed for the preservation of milk and lowered its pH, making it inhospitable to harmful pathogens.[25] In the same period and region, milk curdling via culturing also began, with animal stomach lining, or rennet, added to fermenting milk to make fresh brined cheese (today generally known as feta). Over the centuries, a range of fermented forms of milk developed within the region and spread far beyond, but the East remained the heartland of such practices.

Regional traditions of animal husbandry, along with fermented and cultured milk, would later frame Bulgarian food narratives that looked to reconstruct "national" tradition. Communist-era historians of Bulgarian yogurt look to the ancient Thracians, for example, as the first producers of "Bulgarian yogurt" and *sirene*, a fresh brined (feta-like) cheese.[26] Although Slavs and Proto-Bulgar tribes established themselves in the eastern Balkans only in the seventh century, in the 1960s Bulgarian scholars began to posit that these groups had mixed with the remnants of the ancient Thracians.[27] This theory made its way into food writings, where it was assumed that Bulgarians had inherited not just genetics, but also enduring cultural practices—most notably sheepherding, but also the making of wine, bread, and fermented milk products. If bread, wine, and cheese offered Bulgarians a European pedigree, yogurt provided a milky basis for something more *nationally* distinct in the European context. As communist-era writings on yogurt explored, it was the Proto-Bulgars, a Turkic tribe from Central Asia, who brought another tradition of fermentation, *koumis* (fermented mare's milk), to the region.[28] Eventually *koumis* gave way to the more prevalent practice of making "sour milk" from sheep, goat, cow, and water buffalo milk. In short, Bulgarians claim multiple lines of ancestry and/or territorial continuity with ancient practices of fermented milk, but also the rich culture of animal husbandry in the Eurasian steppes and mountains.

These cultures flourished in the eastern Balkans, a region intersected by mountain ranges and lowland valleys, with the adjacent warm waters of the Black and Aegean Seas. Animal products were always an important element in the local food supply, both for villagers, who kept small numbers of animals, and the transhumance shepherds, who pastured larger herds. These herdsmen have an important place in the historical mythology of yogurt. After the morning milking, local herdsmen would fill bags made from animal skin with fresh milk, which they carried with them throughout the day. As the herdsmen moved, the bags would be shaken, while the bacteria in the bags and the heat of the herdsmen's bodies combined to "sour" the milk, creating a tasty yogurt by lunch.[29] But yogurt and other animal products—especially butter, cheese, wool, and of course meat—were also an important factor in regional trade. Balkan herds supplied food and other products to the webs of towns and larger population centers in the region.

By far the most important period for the development of the regional milk trade was the five centuries of Ottoman rule, from the 1400s to 1908. In consolidating their control over the province of Rumelia, later referred to as "European Turkey," the Ottomans decimated the Slavic feudal nobility, putting into place a flexible and ever-changing system of Turco-Ottoman military elite landlordism. In practice, the Ottoman state owned a large percentage of the land, granting usufruct rights to local peasants, as well as villages and towns. Eventually a de facto system of peasant smallholdings emerged in the mountainous expanses of the eastern Balkans, with larger Turkish estates in the valleys. Animal husbandry was commonly practiced, on a small scale for the most part, especially in the mountainous areas where Slavic Christians tended to settle. This animal-based economy was described in a mid-nineteenth-century essay by Liuben Karavelov, one of Bulgaria's most respected men of letters from the so-called Bulgarian National Revival period (1762–1878). In the mountainous regions, Karavelov noted, "Every peasant had a few sheep, cows, buffalo, [and] goats." But they also had shepherd communes or associations, in which they would

FIGURE 3.2. Young shepherds milk sheep in the fold in the Sofia region, circa 1920s. http://www.lostbulgaria.com/?p=4332.

pool their marked herds, creating a system of grazing and dairy production, with assigned makers of cheese and butter—products that could be sold in towns and cities.[30]

These larger flocks were wintered in the valleys and returned to the mountains for the summer, where they were distributed to their local owners. Some would be driven on the hoof for sale in the more distant city markets of Thessaloniki, Edirne, Izmir, Bursa, and especially Istanbul, the capital of the empire. Local shepherds drove their animals through the Balkan valleys using the Ottoman system of caravanserais, roadside inns located a day's travel apart, where animals and people alike could rest for the night. By the late eighteenth century, Bulgarians were a core element of the armies of itinerant men in the Ottoman animal husbandry trade who left home in March and returned in November. These overland caravans drove herds of largely male animals to Istanbul and other faraway cities to sell for wool and slaughter, while the females were kept behind for continued milk production. As far as trade in milk products was concerned, cheese and butter were generally most suited for trade over long distances, with supply chains for milk products and fresh milk—largely for fermentation—extending from the village to support nearby towns and cities.

Although these techniques allowed for a steadier supply of preserved milk products, milk and other dairy products were not consumed daily by most Bulgarians prior to World War II. For one, their use was limited by the Orthodox Christian tradition of fasting—which generally meant abstaining from meat and/or dairy—for almost half the days of the calendar year. This had a religious as well as economic function; it rationed dairy, so to speak, and it allowed peasants to sell excess milk and buy salt, sugar, oils, and other market needs.[31] There was also a seasonal element to the milk supply: without modern techniques of insemination, the availability of milk was tied to lactation in the spring and summer. Milk was gathered in season and preserved as yogurt, or for longer-term storage as cheese, to be consumed as milk production tapered off. Travelers through the region in this period noted that Bulgarian peasants ate mainly bread, cheese, onion, and garlic. These may well have been the most commonly consumed foods, but they were also *portable* foods that travelers were likely to see peasants eating when they were out and about and near their scattered fields. By the interwar period, more in-depth ethnographic studies noted that a range of dairy products were prevalent in local diets—including yogurt and cheese—supplemented by seasonal vegetables and fruits, as well as legumes.[32] These may have, in part, reflected shifts in diets by that period, but also more longitudinal and in-depth studies of diets, across regions and seasons.

It is worth noting that in contrast to peasants, wealthier Ottoman and post-Ottoman urban dwellers had a higher dietary intake of dairy and other animal products. As a general rule, Ottoman urban households had their own milking animals, which they kept within their high-walled inner courtyards, a common architectural feature of Ottoman cities. But larger Ottoman cities also had *mlekari* (milk guilds), supplied by peasants from surrounding villages. Caravans of oxcarts brought fresh milk and milk products, along with produce, into towns and cities on weekly market days. Oxen—the males of the local water buffalo breed—were the ideal beasts of burden for travel over the mountainous terrain and rutted dirt roads that characterized the Bulgarian countryside well into the interwar period. Travelers reported seeing hundreds of oxcarts "crawling like insects" across the country, bringing wares to market from the villages. Their slow pace allowed peasants to sleep in the back on these distant journeys, for the animals knew the routes well. In her memoirs, Raina Kostentseva describes how turn-of-the-century peasants brought fresh milk and milk products from the nearby villages into Sofia, autonomous Bulgaria's capital as of 1879. The milk was delivered to markets, directly to families, and to local *mlekari*, whose small shops featured boiled fresh milk, their own yogurt, and pastries. Yogurt was also a common Ottoman and post-Ottoman street food, offered by a multiethnic cast of wandering vendors. As Kostentseva describes, yogurt sellers "carried little bowls on trays thrown over their shoulders like a yoke."[33] Clearly milk was an important link between the rural and urban food chains.

Fermented and cultured milk products were also a binding agent in Ottoman cuisine. The notion of a coherent Ottoman cuisine is, of course, somewhat problematic, given the extent and diversity of the empire, with its range of ethnic groups and classes. Diet depended largely on region, season, religion, ethnic group, urban versus rural location, class, occupation, and a number of other factors. Nevertheless, a rich albeit varied Ottoman cuisine developed and evolved, especially in the empire's cosmopolitan urban environments.[34] In the cities of Ottoman and post-Ottoman Bulgaria, a rich food culture percolated through the multiethnic terrain, which included Slavs, Greeks, Armenians, Jews, Turks, Albanians, and Bulgarians.[35] In urban centers consumer taste and food culture were shaped by Ottoman elite culture, itself an evolving amalgam of Eastern and Western influences. Central Asian tradition (including yogurt) became more refined with Persian and Arabic influences, as yogurt came to be featured in a range of dishes, sweet and savory, including soups, noodles, and stewed fruits.[36] Undoubtedly, rich urban Bulgarians were privy to such delights.

But it is unclear the extent to which yogurt was integrated into the cuisine as a whole for nineteenth-century Bulgarians. As noted in past chapters, the famous Bulgarian literary figure Petko Slaveikov authored the first cookbook in the Bulgarian language, which was published in 1870.[37] His *Gotvarska kniga* includes various basic culinary techniques for storing, preserving, and preparing foods, including yogurt. To paraphrase, Slaveikov advises readers to skim off the cream, heat the milk until it is almost boiling, pull it off the stove, and let it cool until you can tolerate the heat with your finger; then pour starter (a spoonful of yogurt) into the pot, lightly cover it to allow steam to escape, wrap it in a cloth, and let it sit for a few hours. "Over time," he adds, "the milk will become thick, and in terms of flavor—sweet and delightful," noting that "in warm weather it will last for 3 days, and in cool weather a whole week"—unrefrigerated, of course.[38] The process of making yogurt was important enough to be featured, but yogurt as an *ingredient* was notably absent from the dishes in Slaveikov's book. This was in stark contrast, as we will see, to twentieth-century Bulgarian cookbooks, in which yogurt's milky imprint on Bulgarian cuisine vastly expanded. A number of developments in the post-Ottoman period would set the stage for this shift.

The establishment of the Bulgarian principality in 1878 in the wake of the Russo-Turkish War brought about a radical change in the socioeconomic order—creating new possibilities but also vulnerabilities. The effective loss of open trade in Ottoman markets led to a major scaling back of animal husbandry, which was also crippled by the chaos of the Russo-Turkish War. The borders of the new Bulgarian principality choked off transhumance winter pasturing and adversely affected caravan trade with adjacent Ottoman cities. Furthermore, as a segment of new Bulgarian elites urbanized and accumulated wealth, an increasingly heavy tax burden, waves of economic crisis, and new levels of peasant debt contributed to greater wealth disparity.[39] Disillusionment washed over all segments of Bulgarian society, not just peasants and the burgeoning urban lower classes. Western economic and cultural penetration accelerated with quickening globalization and presented a new threat—in some senses even greater than the Ottoman one. If the West as a model had a certain allure for many Bulgarians, for others it was seen as a source of condescension and exploitation of "small nations" like Bulgaria. It was to blame for continued and deepening Bulgarian "backwardness," and also for some of the visible flattening effects of "civilization," which were particularly stark in Bulgarian cities.[40]

The Europeanization of urban culture, accompanied by successive waves of Muslim and later Greek emigration, meant that Bulgarian cities became less recognizable to many of their longtime inhabitants. This process arguably

opened the way for a selective embrace of the Ottoman past and its legacies as raw materials, nay ingredients, for local and national distinction.
In food as in other realms, aspects of Ottoman-era culture were readily mixed
with European influences and then recast as Bulgarian to provide a bulwark
against the homogenizing onslaught of Western culture. This was a drawn-
out process, however, and it was not until the interwar period that any cookbook made claim to codification of a national cuisine. Arguably, in the first
half of the twentieth century the impulse to create a national *cuisine* was still
largely overshadowed by growing concerns with biochemical deficiencies in
the national (and especially peasant) *diet*. While protein was more important than probiotics in such writings, yogurt provided both. With Mechnikov
(and Grigorov) as inspiration, yogurt found a place in new ponderings about
Bulgarian health and longevity, as well as a national distinction.

FIGURE 3.3.   Milkman in Sofia, circa 1920s. http://www.lostbulgaria.com/?p=2467.

As noted above, in parts of the West, consumption of Bulgarian yogurt had spread among the upper and middle classes as a supplement to an already protein- and dairy-rich diet. It was of interest as an exotic, authentic, even Oriental novelty with enticing health benefits—real as well as imagined.[41] In Bulgaria, in contrast, yogurt was never a novelty or trend, let alone exotic or controversial. Instead, changes in dairy production—and an expanded advocacy for yogurt consumption—were tied to interwar and postwar modernizing impulses, and a new awareness and concern for the "protein gap" between Bulgaria and the West, as well as between social strata within Bulgaria. As noted in the preceding chapters, the near starvation of Bulgarians in the last years of World War I provoked local scholars to launch studies that charted the geography of the Bulgarian diet in fine detail. Local thinkers and newly trained food scientists were influenced by studies and narratives coming from Western scholarship and institutions like the League of Nations, which commissioned studies of food patterns across Europe.[42]

Armed with newly produced international data and food science, they looked with alarm at the meager bread-based diets of the Bulgarian peasantry (still the bulk of the population) and the urban poor. A lack of protein was the key finding, and for many the Western meat-rich diet remained the ideal. For others, however, a diet with increased dairy (and eggs) was far more realistic. Dairy was a more affordable, and in a sense renewable, source of protein—as well as fats, vitamins, and minerals—for the hungry masses.[43] For prominent food scientist Asen Zlatarov, the issue was both sustainability and health. As a vocal vegetarian, Zlatarov focused on milk products as an important source of protein, along with legumes and whole grains. He and a range of other vegetarian advocates—discussed in chapter 2—were eager to see yogurt, cheese, and other forms of dairy become a staple of the Bulgarian diet.

In and beyond interwar vegetarian communities, dairy became an ever more prominent hallmark of Bulgarian cuisine. New cookbooks and the recipes that began to appear in women's magazines such as *Vestnik na zhenata* (Journal for women) featured yogurt and other dairy products as ingredients in a range of newly codified dishes. *Nova gotvarska kniga* (New cookbook) was the first cookbook to offer and promote avowedly "Bulgarian" dishes that were at once "tasty, nutritious, and economical."[44] Unlike Slaveikov's earlier *Gotvarska kniga*, it contained no instructions for making yogurt, perhaps because the process would have been widely known, and also because yogurt was readily available for purchase. But also unlike *Gotvarska kniga*, the new edition featured yogurt as a common *ingredient* in a variety of dishes, including hot and cold soups and *tarator*, which is featured there not as a cold soup (as one would find it today) but as a "salad" of yogurt, cucumber, garlic, salt, oil, and vinegar. Interestingly, in a section on *"postni i vegetarianski gozbi"* (vegan-fasting and vegetarian dishes), one finds both dairy-free and dairy-laden options, including (my personal favorite) fried zucchini swimming in yogurt.[45] With a spoonful here and a ladleful there, yogurt was ever more thoroughly mixed into Bulgaria's evolving foodways.

As Bulgarians looked to Europe for dietary models, they also began to look to milk science and technological methods and standards in dairy production. Changes were admittedly modest by Western or postwar Bulgarian standards. Still, new milk collection points, dairy stations, veterinary stations, and milk science laboratories were established, in concert with efforts to modernize food chains, train new experts, and create dairy industry regulations and standards.[46] A new crop of Bulgarian policy makers, agronomists, biologists, chemists, and dairymen contributed to efforts to build local infrastructure for dairy farming and yogurt production. An important element of such efforts was tied directly to bacteriology, namely an understanding of both sterilization and fermentation. Asen Kantardzhiev's 1930 book *Mlekarski narechnik* (Milk manual), for example, is devoted almost entirely to laying out the necessity of and methods for eliminating harmful bacteria, but also for culturing the *healthy* bacteria in yogurt—not just *Bacillus bulgaricus* but also *Streptococcus thermophilus* (the other common bacteria in Bulgarian yogurt).[47]

In the shadow of Western milk science, as developed at the Pasteur Institute and elsewhere, traditional methods of yogurt production came under scrutiny. For even Mechnikov, in spite of his love for Bulgarian yogurt, was critical of the inconsistent and unhygienic methods used to produce it. In rural homes, it was noted, a variety of "wild" and not necessarily beneficial bacterial cultures bloomed—not just *Bacillus bulgaricus* and *Streptococcus thermophilus*. Yogurt was most commonly produced by rural women, who warmed milk to a certain temperature—dipping a finger in to test it—and then added some starter (*maya*) from the last batch of yogurt. Clay vessels of yogurt were kept near the fire, covered with wool or other materials, which in some cases were dipped into the yogurt to save some of the starter in a dried form. At the Pasteur Institute, Mechnikov had been keen on eliminating "wild" strains and creating a cultivated and "clear" formula for use in yogurt.[48] This became the standard in much of Western Europe and the United States, also appealing to many of the new dairymen of interwar Bulgaria.[49] And yet such aspirations and advances in interwar milk "industrialization" were extremely limited; they primarily took the form of an increased number of urban craft dairies.[50] Meanwhile, most yogurt was still produced on private farms and hence retained many of its "wild" properties. Tradition, it seems, remained the order of the day, given the still limited resources and urban markets for real change in dairy production.

But yogurt tradition continued to be revered, even as it became entwined with science and history in interwar milk narratives. In Bulgaria, in contrast to Western Europe, emergent narratives of milk scientists and dairymen singled out yogurt as having important *national* as well as nutritional significance. As noted by Kantardzhiev, for example, yogurt was by far the best way to consume milk, as it made fresh milk easier to digest, was not a source of hygienic concerns like fresh milk, and, unlike cheese, was more healthful—even prophylactic—against a range of diseases.[51] But yogurt was also a "national food" and therefore deserving of "special attention." For Kantardzhiev, the bacteria in Bulgarian yogurt were a source of national pride, the best in terms of taste, but also biochemical constitution, as "proven" in a cited Swiss study. In language hauntingly evocative of racial science, he claimed that this was most likely the result of a kind of "natural selection" that had occurred within Bulgaria, where local production had weeded out the less desirable strains.[52] Similarly, K. Popdimitrov, another advocate of modern dairying, described the ways in which Bulgarian yogurt had been embedded in "Proto-Bulgarian" culture, tradition, and nutrition since ancient times—that is, predating the Ottoman conquest.[53] From history Popdimitrov pivoted to microorganisms, explaining how they had become a cornerstone of bioscience in the previous one hundred years, but also played an important role in food production—not just yogurt, but also beer, wine, spirits, vinegar, and bread.[54] His subsequent detailing of the microbiology of Bulgarian yogurt was intertwined with the names Pasteur, Mechnikov, and Stamenov.[55] As M. Iordanov explained in the foreword to Popdimitrov's book *Bŭlgarsko kiselo mliako* (Bulgarian yogurt), yogurt—not just any yogurt, but *Bulgarian* yogurt—had attracted international attention for its healthful properties among "scholars, economists, sociologists, and the press."[56] The state and "every intelligent Bulgarian," he elaborated, should work to increase the consumption of yogurt for the health of the nation, not just because of the yogurt's microflora but also for the proteins it contained.[57]

With the Bulgarian yogurt fad slowly petering out in the West, a small but growing number of Bulgarian scholars began to take notice of the place of Bulgarian yogurt in Mechnikov's work. It was not just dairymen, but also social scientists who undertook their own studies of local centenarians.[58] Interestingly, one such scholar cast doubt on Mechnikov's numbers: Vladimir Partŭchev had estimated that there were 3,800 centenarians in Bulgaria, but locals were able to verify only 158. Nonetheless, he noted that Bulgaria was second in the world (after Guatemala) in centenarians as a percentage of the population.[59] While downplaying the role of yogurt in his findings, Partüchev listed it as one of many foods that most centenarians consumed, but noted that food was just one piece of the puzzle.[60] These centenarians, the bulk of whom were ethnic Bulgarians, would be a serious subject of study for years to come, living evidence that rural Bulgarian life and food-ways could mean a longer life span. Perhaps it was not their intent, but this fed into the larger tendency among intellectuals from the late nineteenth and early twentieth centuries to cast aspersions on the ills of "civilization," which they viewed as festering in the Bulgarian city and coming from the distant West. As an antidote, intellectuals on the right and left embraced the "folk," their ways and wisdom, which of course included yogurt.[61]

Mechnikov's "milk science" made a global splash, with a delayed and quite different effect on Bulgarian soil. In spite of the growing number of scientific studies of agriculture and animal husbandry, there were no major changes in modes of milk processing in pre–World War II Bulgaria. For the time being, Bulgaria remained completely outside the boundaries of the Western transition to "modern" milk, predicated on pasteurization, on the pure and the fresh. While such modern practices were not entirely absent, they were suspect, for they would mean the literal killing of the Bulgarian fountain of youth, the *Bacillus bulgaricus* that made milk curdle and become thick and smooth, supple and sour.

Back to the Future

In the years that followed the Second World War, the processes that had begun in the interwar period would markedly accelerate under a new set of state-driven imperatives and possibilities. As the new regime took control over the economy with a new kind of will and zeal, people and their bodies were also slated for a kind of restructuring. Shifts in agricultural and dietary practices went hand in hand as the new state ministries sought to rationalize food production and consumption. A growing preoccupation with "rational consumption" precipitated the counting and charting of every gram of food as well as its caloric and nutritional information. Socialist states, including Bulgaria, encouraged and directed their populations to increase their con-sumption of certain foods, striving to match or surpass the West, where caloric intake and nutritional statistics were also closely tracked. As noted in the last chapter, protein was an important part of the Cold War contest, in which bodily fortification was a marker of systemic success. Along with meat, milk products were a cornerstone of the prescribed dietary restructuring.

The whole gamut of dairy products was an important ingredient in making Bulgarians modern, in building socialism. Yogurt, however, held a privileged place in these processes—with its powerful and entwined scientific, historical, and national mythologies.

In socialist Bulgaria, the high nutritional priority given to increasing the consumption of milk products necessitated the development of a full-blown dairy industry. As a 1949 dairy industry handbook clearly and simply states, "With the building of socialism, we need to provide ample and cheap milk."[62] The hope and expectation was that milk products would be a central part of the *daily* diet for all Bulgarians, which would require a major set of shifts in production and consumption practices. Up until this point, milk products had been consumed irregularly, based on what the communist state saw as "irrational" religious practices associated with vegan fasting. Under strict Orthodox rules, consumption of animal products was prohibited, at least in theory, on almost half the days of the year, including Wednesdays and Fridays, as well as for longer periods during Lent and Holy Week. But low levels of dairy intake were also governed by the natural seasonality of milk and the poverty of large segments of the population, who interwar sources had shown were living primarily on grains. Socialist restructuring called for an end to irrational religious prohibitions on meat and dairy, as well as for a redistributive leveling of incomes and a higher standard of living for all. More milk products—not just yogurt, but also cheese and fresh milk—would provide rich and delicious daily proof of the power of the new regime to provide for its populations. Like meat, dairy was seen as a critical human *fuel* that helped propel bodies to perform agricultural and industrial labor. Meat, as explored in chapter 2, was an important factor in the move to provide increased protein, but dairy was at least equally important, as it was much more practical and affordable.

The immediate postwar period brought a rash of publications on modern industrialized milk practices, indicative of the fact that milk for the masses was a pressing priority.[63] In the eyes of communist modernizers, the inherited state of Bulgarian dairy production was miserably backward and would need a total restructuring. As described by dairy specialist Nikola Dimov in 1949, for example, "Dairying in Bulgaria is backward in all respects. In small, individual, and uncoordinated agricultural pursuits, it is hard to increase or improve milk production."[64] Most peasant households had only one cow, plus a handful of sheep and goats; the latter produced far richer but smaller yields of milk, which was most often made into yogurt and cheese. As the goal was to dramatically increase milk production, a whole range of changes would be necessary at all stages of the farm-to-table supply chain to make the newly milk-rich diet a reality. The collectivization of agriculture, at first voluntary and then forced (1949–51), consolidated large numbers of small-scale farms into more efficient state-controlled operations. In 1949 the state also reopened and began to revamp Serdika, the first state-run dairy, in Sofia. It became one of a string of state-run dairies that took over the production and distribution of yogurt, cheese, and other dairy products.

Science, technology, and shifts in animal stock were slowly but surely brought to bear on Bulgarian production and supply chains. A legion of milk scientists followed in the footsteps of the oft-cited Stamen Grigorov and Ilya Mechnikov, who were lauded as the founding fathers—the Marx and Lenin—of Bulgarian milk science. These scientists worked on concrete problems in the new Bulgarian milk industry at all stages of production, including the ongoing issue of increased supply, which remained a perceived problem for decades. As late as 1960, a book on dairy production noted with concern, Bulgarians were still in the category of "low milk production and consumption" in comparison to the rest of the world, with the inhabitants of Sofia consuming a mere 0.15 liters a day, in comparison with Berlin (0.45) and Prague (0.5).[65] The Sofia figure presumably referred to yogurt, and not milk per se, which was consumed far less in its fresh form than in its fermented form. In Bulgaria in the 1950s and 1960s, mass-produced yogurt was distributed "on tap" in cities, towns, and even some villages. That is to say, people brought their own five- or ten-liter metal containers to be refilled at state-run outlets.[66] By the 1960s, metal containers had given way to returnable small glass jars, which were in turn replaced in the 1970s (at least in Sofia) by plastic containers with foil lids.[67] There were continuous efforts through the period to study and improve all stages of the production process, as well as to upgrade supply and distribution chains.

Grand leaps in production were facilitated in large part by a dramatic transformation in the *kind* of milk people consumed. In the course of the communist period, Bulgarians moved away from yogurt made from sheep's milk in favor of cow's milk. As the milk animals that produce the largest volume of milk and also best lend themselves to mechanical milking, cows had already come to dominate the milk industry in the West because of higher yields, and perhaps also tradition and taste. In Bulgaria, in contrast, cows had made only modest inroads into local dairying and had been traditionally utilized more for labor than for food, with sheep providing the favored form of milk and hence yogurt.[68] Even milk modernizers from the interwar period were less keen on cow's milk, which was deemed inferior in taste and consistency for yogurt production.[69] Under communism, sheep husbandry and milk production were largely reoriented to cheese production, and sheep's milk production remained relatively flat in terms of quantity throughout the period. Bovine dairy output, in contrast, increased from 450,000 liters in 1939 to 2,259,000 liters in 1974.[70]

As production increased, new structures undermined home production of yogurt; as people joined collective farms, populations urbanized. But it is worth noting that small-scale animal husbandry and fermented-milk production continued in the smaller towns and mountainous regions. Collectivization did not preclude the practice of maintaining small plots and a small number of animals for personal use. This allowed rural inhabitants to continue to produce and consume sheep, goat, and buffalo milk, yogurt, and cheese. And since most urbanizing Bulgarians retained ties to the village, such sources of milk were never far away. Village diets did not change dramatically among the older generations or, especially, among certain minority populations, who were not part of the mass urbanization of the period. To be sure, urbanites—who brought their children back to the village to stay for the summer—brought certain urban products, tastes, and practices back with them. But fresh village-grown ingredients retained a privileged place in the Bulgarian diet and culinary imagination. Peasants also continued to bring milk and milk products into town from the villages to sell on street corners and in markets, or to provide to family members. This was a source of income, and often a necessity, given the shortages and inconsistent milk supply lines in Bulgarian cities. Undoubtedly the flavor of this *real* Bulgarian yogurt hung like a cloud over the milk industry, for which change did not come without cost.

If milk quantity continued to be a preoccupation throughout the period, the 1950s saw efforts to increase *quality* become more pronounced across Bulgarian state planning structures. This was an element of what scholars have deemed the post-Stalinist "consumer turn," but it was also part of efforts by the communist state to satisfy the desires of a new urban consuming citizenry with higher expectations. By the 1960s, there were competing concerns within the industry—that its mechanization had not gone far enough, but also that it had had deleterious effects on yogurt's flavor and nutritional qualities. The remedy was intensified research and development and a return to the villages, to seek out the cultures that had been lost in the mechanization process. In the course of the 1960s and 1970s, a number of institutes throughout the country began gathering cultures, then standardizing and patenting a "Bulgarian yogurt" bacterial starter.[71] In this period, socialist producers were extremely careful about preserving the flavor and nutritional benefits, even if that meant less efficient processing practices. When large-scale boilers were acquired from France in the 1970s, for example, Bulgarian yogurt makers were appalled by the "lack of flavor" that resulted from their use.[72] Bulgaria's milk professionals became increasingly preoccupied with preserving (or reviving) the health benefits of Bulgarian yogurt through their scientific studies.

Bulgarian socialist sources looked to but also greatly expanded on Mechnikov's turn-of-the-century "orthobiotic" (i.e., probiotic) claims regarding Bulgarian yogurt. The focus was increasingly on yogurt's bioactive *Bacillus bulgaricus*, which was lauded for its miraculous digestive and healing properties, as well as its ability to decrease toxins from the "rotting process in the intestines," treat diarrhea, relieve headaches, improve one's psychological state, boost appetite, lower a fever, and help in recovery after surgery.[73] Researchers also claimed that yogurt reduced cholesterol. Cholesterol was certainly a concern in Bulgarian nutritional science, and there was no lack of nutritional experts who advised limiting "animal fats," if only for those with arteriosclerosis. But in most cases, yogurt was not lumped in with the category of "animal fats." On the contrary, it was offered as an alternative to meats. Not only did many sources assert that full-fat yogurt was not contributing to the cholesterol problem—they also offered studies that "proved" that consuming it actually lowered cholesterol, and that it was a "prophylactic" food used in most hospitals for a range of illnesses.[74] While some might see such conclusions as spurious, it would be remiss not to note that an entire school of recent nutritional science in the West has embraced the "heart healthy" aspects not only of "probiotics" but also of full-fat, grass-fed dairy.

Interestingly, Mechnikov's theory on Bulgarian yogurt (which had always been controversial) slowly lost its luster in much of the West. Mechnikov's theories had been questioned from the very beginning, but this was particularly the case after his death in 1916. By the interwar period, various studies seemed to disprove his ideas on yogurt and the importance of the famous microbes, or sidelined their usefulness to having minor therapeutic value.[75] Nevertheless, in the interwar and postwar periods, yogurt continued to be consumed and massively gained in popularity in France, parts of Western Europe, and the United States—in part because of Danone. Danone continued to ride the wave of the Bulgarian yogurt craze in Paris until World War II drove the family and the company to American shores. Under the label Dannon, the firm expanded its presence there, but it also came back to Europe as a multinational corporation in the postwar period. By then Danone had retooled its product and marketing—shifting away from Bulgarian bacteria and instead adding fruit and sugar to its yogurt, which it now presented as a posh dessert-like snack. In the US market, all references to its Balkan origins were dropped; if anything, yogurt was seen as French, especially when Yoplait entered the market in the 1970s. The postwar version of American yogurt also tended to lack "live cultures" because of post-fermentation heating, and included a range of thickeners and additives that were used to extend its shelf life.[76] It was only local brands of "hippy" yogurt (for lack of a better term) that continued to produce a primarily "plain" variety and focus on live cultures. It was not until much later, in the 2000s, in the fading afterglow of the Cold War, that the term and concept of "probiotics" went mainstream.

On the east side of the Iron Curtain, in contrast, the belief in probiotics (although that term was not in use at the time) not only survived but reigned supreme, especially in Bulgaria—the homeland of Mechnikov's anointed variety of the invisible bacterium. In socialist Bulgaria, Mechnikov's science was gospel, inspiring a rash of writings on "long life."[77] This seemed to be a matter of state importance. As noted in one study on the subject, communist leader Todor Zhivkov proclaimed in 1961, "The struggle for longevity is now being waged alongside the struggle to build socialism." Bulgarian longevity was a measure of socialist progress—its advances in health care, hygiene, and "feeding the public."[78] Mechnikov's "orthobiotics," which rested on the "microflora" of Bulgarian yogurt, was repeatedly presented as a kind of cure for aging.[79] Again and again, Bulgarian sources—without exploring counterclaims—noted that decades of research had only confirmed Mechnikov's ideas in relation to the biochemistry of yogurt and its healthful properties.[80] But this connection of long life both to "progress" and to Mechnikov's ideas presented certain contradictions. After all, if it was "progress" that had created the conditions for Bulgarian longevity, then why were all the centenarians in remote villages, far from the scientific advances that had been introduced into Bulgarian urban life? It was these very contradictions that opened the door for an alternative path to progress in the realm of food. In a sense, yogurt, as sanctified by Mechnikov, allowed for Bulgarian milk scientists and historians to embrace the wisdom of the past and question the effects of industrialized milk and yogurt.

Without a doubt, the yogurt-lined path to Bulgaria's future ran circuitously through the past. The history of milk fermentation played a critical role in yogurt's elaborated mythology under socialism. This history was repeatedly detailed in scholarly and popular works on yogurt and dairying. As in the interwar period, the Proto-Bulgars, with their tradition of fermented mare's milk, were often cited. But now they were upstaged by the notion of ancient Thrace as the original site of milk fermentation and (soft-cheese) culturing. Archaeological finds—namely, Thracian burial mounds discovered in Bulgaria in the 1960s and 1970s—provided evidence of the tools and vessels used for such purposes. In tandem with new theories claiming that Bulgarians were descendants of the Thracian peoples, the notion of Bulgaria as the ancient birthplace of yogurt took hold. Yogurt was dislodged, at least in part, from Turco-Ottoman tradition and cast as both ancient and essentially Bulgarian.

This preoccupation with the ancient past might seem counter to a forward-looking Marxist regime, but Marxism was also deeply concerned with historical teleology and material culture. The Thracian theory offered Bulgaria a bona fide pedigree, a niche in the "cradle of Western civilization," and the material basis to make claims about an ancient *European* culinary tradition. According to a source from the period, the yogurt tradition had lasted since ancient times because of the "unexplainable and mysterious" benefits of this "perfect food."[81]

Recipes for the Future

Popularizing the findings of milk research—biochemical as well as historical—served to encourage the consumption of yogurt and other milk products in socialist Bulgaria. Scholars' findings were published in stand-alone pamphlets, books, and articles, but they also provided a narrative thread in cookbooks, women's magazines, and other sources that promoted a new food culture in socialist Bulgaria. Beginning in the Khrushchev era, the Soviet Union gave Bloc states the green light to focus on providing a more robust and varied consumer culture, in which food was central. In the Soviet Union and across the Bloc, consumer offerings were also an important factor in bolstering legitimacy domestically. If deepening and extending Bulgaria's milk-based cuisine started as an efficient way to properly feed the population, it evolved into a key component of creating a culinary consumer culture, as well as a national cuisine.

As noted in chapter 2, women had a special role in this process, as they were expected to purchase, preserve, and prepare food in a whole array of new ways. From the earliest years, cookbooks were an important medium for the socialist regime to propagate nutritional science among women, to teach them how to feed their family a "rational diet." The earliest socialist-era cookbooks devoted an inordinate amount of space to the science of food as part of their drive to increase calories (and especially protein) and hence productivity and wellness.[82] The state needed knowledgeable and willing consumers, primarily women, to translate their milk science into daily consumer practice. The 1955 cookbook *Nasha kukhnia* (Our cuisine), for example, contains a section on milk that lauds milk, whether of cows, buffalo, sheep, or goats, as a "complete food." Yogurt, in particular, is celebrated for its many nutritional gifts as well as its cultural meaning: "Through the fermentation of milk, one can prepare yogurt. This is a very valuable food, rich in food substances, especially lactic acid [fermented lactose created by the bacteria in yogurt], that help with the regulation of digestion. Bulgarian yogurt is the national food."[83] In spite of this invoking of yogurt as "national," *Nasha kukhnia* was not a self-proclaimed "national" cookbook. Rather, it offered techniques for how to cook "our cuisine."

"Our" was open to interpretation, especially given that food culture was in many respects shared by Bulgaria's majority and minority populations, primarily Turks, Pomaks (Bulgarian-speaking Muslims), Roma, and Sephardic Jews, but also smaller numbers of Greeks, Romanians, and Gagauz (Turkish-speaking Christians). As has been well documented, in the early years of communism the differences of such groups were embraced and encouraged, following Lenin's dictum "National in form, socialist in content." This is reflected in *Nasha kukhnia*, where the authors note, "Our cuisine is close to Oriental, and also to Russian. The long years of the Turkish yoke and Russian influence after liberation have left significant vestiges in our cuisine."[84] But beyond a nod to spices and rice, they are not clear on which influences or dishes were "Oriental." Instead, the plethora of techniques and recipes they provide constitute a mishmash, with "Russian" as well as familiar local recipes. Yogurt is far more prominently featured than in interwar cookbooks—not how to make it, but as a key ingredient in a range of recipes, including four different kinds of *tarator* (the familiar cold yogurt soup), summer borscht, green bean stew, and salads with cucumber and radish.[85] Still, the focus is not on culture or the articulation of a national cuisine, but rather on nutritional science and proper culinary technique—or, for yogurt, on protein and probiotics.

This makes it all the more interesting that yogurt was singled out as "national." It foreshadowed the socialist state's increasingly open deployment of nationalism after 1956, a period in which Bulgarian minorities, with the exception of Jews, were targeted for gradual and total assimilation. As part of this process, all visible signs of Muslim difference were deemed backward and retrograde, a barrier to progress in a modern, "European" state.[86] Almost all that was "Oriental" came under gradual and increasing attack in virtually every realm of socialist culture. All, that is, but food.

By 1978, Liuben Petrov's *Bŭlgarska natsionalna kukhnia* (Bulgarian national cuisine) provided the first avowedly *national* cookbook of the socialist era. In the introduction, Petrov detailed a 1972 meeting organized in the Bulgarian city of Plovdiv by the Ministry of Internal Trade and a cooperative union of cooks from fourteen different culinary institutes. Over two days, the participants prepared some sixty dishes, with the goal of "reviving" Bulgarian "traditional" cuisine.[87] Petrov explained that participants had gone to "the people" to gather recipes from "grandmothers, old chefs, and housewives." In addition, he had consulted Petko Slaveikov's 1870 "national" cookbook from Istanbul. This was a project in which the Ottoman past was not easily discarded, especially as Slaveikov wrote his volume in the Ottoman capital. Regional foods, like yogurt, were certainly coded as "Bulgarian," but as Petrov also notes, the most important influences on Bulgarian food were Greek and Turkish.[88]

This and other sources provided ballast to an emergent Bulgarian national cuisine, awash in yogurt.[89] Yogurt emerged as an important constant in Bulgarian national cuisine, something that could stand alone, be topped with something sweet (like honey or stewed fruits) as dessert, or "bind" other ingredients.[90] Literally hundreds of recipes either included yogurt as a major ingredient or were punctuated by the refrain "serve with yogurt." It was used in salads (green, potato, eggplant, roasted red pepper), usually with garlic and herbs; and in soups (bean, potato, chicken, fish), in which it was a thickener. It was also used as a sauce (often with garlic) for a range of roasted vegetable and meat dishes. In a few recipes, yogurt took center stage. Here, for example, is the recipe for *tarator*, a popular summer delight, from *Kniga za vseki den*, which recommends daily consumption of yogurt:

> Beat 3 cups of yogurt. Add 1–1½ cups of cold water, salt to taste, 2–3 teaspoons oil, and 1 diced cucumber. If you like, you can add 1–2 teaspoons vinegar and 3–4 cloves of garlic.[91]

In an unusual twist, the recipe suggests sprinkling each bowl with diced hardboiled eggs mixed with black pepper, chopped dill, and parsley. Later recipes more commonly include the option of crushed walnuts. Yogurt is also a thickener in hot soups (pea, mixed vegetable, and leek), the base for garlicky sauces to dollop on roasted red peppers or fried eggplant, and an ingredient in baked goods. It was also mixed with cold water in the summer for drinking, in the still popular yogurt drink *airan*. The yogurt-laden recipes in *Kniga za vseki den* and other cookbooks from the period offered plenty of muscle-building protein and gut-populating flora. In this way, yogurt would allow Bulgarians to not only "build socialism" but live to see the coming of communism, à la Mechnikov's oft-cited theory of long life.

But the significance of Bulgarian yogurt went beyond simply fortifying local diets and offering a vehicle for flavor. The stakes were much higher, as food became a factor in attracting and feeding the millions of tourists who came to Bulgaria each year by the 1960s, primarily from the Bloc but also from the capitalist world. In the Bulgarian food and tourist industry, professionals were well aware that foreigners were attracted to a cuisine that, as *Bŭlgarska natsionalna kukhnia* asserted, "combined the spice of the East with the subtlety of European national specialties."[92] One of this cookbook's stated goals, in fact, was the integration of the "national cuisine" into public food offerings, like cafeterias and restaurants.[93] The impetus, then, was not just to train women to create nutritionally balanced and nationally attuned domestic utopias but also to flavor and color the newly blossoming domestic and foreign tourist industry with its accompanying restaurant offerings.

An interesting artifact from Bulgaria's tourist industry from this period is a glossy-paged culinary travelogue and cookbook that was published in English, German, and French editions, titled *Bulgarian Temptations: 33 Illustrated Culinary Journeys with Recipes.*[94] In it the author and food enthusiast Emil Markov maps the Bulgarian culinary landscape through forays into its history and local variation. He paints a layered (if not seductive) picture of Bulgarian delicacies and their myriad inspirations, openly acknowledging that "the influence of the Oriental is indisputable."[95] In a fascinating alternative to Bulgaria's gleaming resort-city restaurants, the book gives particular attention to simpler culinary pleasures, small guesthouses, and home cooking, highlighting local ingredients. With respect to yogurt, Markov notes that it is "typical of Bulgarian national cuisine," playing both a culinary and a "curative" role.[96] And then there are the recipes, so many of them drowned in yogurt. Yogurt is used to bake his "milk kebab of lamb" and as a thickener in many of his soups, for example his brain soup:

> 500g calves' brains, veal stock made from 100g veal, vegetables and spices, to a total of 1200g, 50g butter, 20g flour, 2 eggs, 100g yoghurt, 100g fried bread. Place the brains in cold water until you can remove the skin, wash well, add the well-browned flour and pour the stock over them. When the soup has come to a boil bind with the egg beaten up in the yogurt, and flavour with black pepper and salt. Serve with diced fried bread.[97]

Shkembe chorba (tripe soup) is a much more commonly found version of this yogurty stew—and, as Markov notes, a known hangover cure. Yogurt, as he describes, is also used in baking, in pastries from the Bulgarian region of Thrace like *banitsa* (similar to *biurek* in the former Yugoslavia, or to Greek spanakopita). It is also on menus in Thrace and across the country, as in the tavern in Klokonitsa that served its guests juicy meatballs (*kebabche*) and aromatic grilled sausages, salads, and ice-cold *tarator* made from ewe's-milk yogurt and cucumbers.[98]

But by this period, inviting foreigners to the Bulgarian table to partake of the country's famous yogurt was not enough. Among Bulgarian milk science and industry professionals, ambitious efforts were under way to "conquer the world" with Bulgarian yogurt.[99] It was difficult to export actual jars of yogurt, so the focus was on licensing and exporting Bulgarian yogurt starter—a mix of *Bacillus bulgaricus* and *Streptococcus thermophilus*. By the mid-1970s, the Bulgarian milk industry had begun to openly peddle yogurt to Western European and American companies like Dannon (the US version of Danone). While most of their direct export possibilities were tangled in a web of impossible regulations, they did manage to license and brand "Bulgarian yogurt." Bulgaria's biggest trade coup in the period was its contract to provide licensing and "know-how" to the Japanese firm Meiji Milk. With the direct cooperation of Bulgarian milk industry scientists, Meiji Milk began to produce "real Bulgarian yogurt" on its own soil. Released to market in 1973, it quickly became a commercial success, using the Mechnikov-inspired slogan "24 million Japanese that consume Bulgarian yogurt will live to be 100 years old."[100] In both name and biochemical form, Bulgarian yogurt was back on the global market—at least in a limited way.

But as with other sectors of the economy, the collapse of communism brought a dramatic collapse of the dairy industry. Between 1988 and 1994, total milk production was cut almost in half, back down to the 1961 level; in spite of ups and downs since then, production totals in 2018 were even lower than in 1994 (and 1961). This at first was due in large part to the chaos of the transition and de-collectivization and privatization. But since 2008, EU regulations have caused a large number of local "unhygienic" dairies to close, also curbing the informal circulation of raw milk.[101] Since 2002, Bulgaria has continued to export select milk products (mostly cheese), but it has also become a major importer of milk products and other foodstuffs and host to multinational corporations like Danone that have set up shop on Bulgarian soil.[102] Post-socialist Bulgaria has gradually been flooded with foreign products. Danone, whose yogurt appeared on Bulgarian shelves and which eventually set up its own dairies in the 1990s, now has a 30 percent market share.[103]

This raises the issue of the effects of post-socialist globalization on Bulgarian food. Not surprisingly, debates on the nature and future of Bulgarian yogurt have accompanied the transition.[104] As Nevyana Kostantinova lamented in her *Call for a Future for the Heart*, "I am looking for meaning in real things. For example in Bulgarian *kiselo mliako* (sour milk). Danone has appeared in our lives. YES, it is *kiselo mliako*. BUT NOT real Bulgarian *kiselo mliako*."[105] This has been a cause for concern among Bulgarians, for whom their country's yogurt mythology is stronger than ever. The name and glories of Bulgarian yogurt can now be heard across the web-o-sphere, with a new and growing intensity in the face of encroaching multinational corporations.[106] Locally produced yogurt persists in rural areas, and there is an expanding market for organic local yogurt in Bulgarian cities. But to the dismay of Bulgarians, "real" Bulgarian yogurt seems to be almost an endangered species at home (at least in the cities), even as it is overshadowed by "Greek yogurt" abroad.

Interestingly, in the 2000s, yogurt has carved out an extensive niche for itself in the exploding global food scene. Probiotics seem to have crossed the Iron Curtain or been elevated from hippy food to the mainstream in a rapidly evolving global food culture. Americans have begun to believe in the science of bacterial gut health, but they also look for pastoral (and again, Balkan) authenticity in the face of a thoroughly modernized food system. This might explain the sudden boom in demand for "Greek" yogurt, which dominates the yogurt aisle in US supermarkets today, much to the chagrin of watchful Bulgarians.[107] The Greek yogurt fad began in 2007, when the small dairy start-up company Chobani, based in New York State, released a thicker, strained variety of yogurt onto the American market. So-called Greek yogurt was such a success that within five years the small company was worth billions in the US market, second only to Yoplait and Dannon. Chobani, ironically run by a *Turkish* (actually ethnic Kurdish) entrepreneur, Hamdi Ulukaya, was clearly in the right place at the right time, as his "Greek" yogurt rode the wave of trending high-protein diets, but also probiotics. Chobani propelled a new market that led Danone and other competitors to follow suit, so that "Greek yogurt" now accounts for roughly half of all the yogurt consumed in the United States.[108]

Greek yogurt's popularity in the 2010s is no surprise. It has the potential to appeal to the brave new world of science, and an imagined pastoral (and distant) past, even if Chobani was sued (to no avail) for selling a product that was not made in Greece or by Greeks.[109] The newly spawned "bubble" in Greek yogurt demand harks back to an earlier time, at the turn of the twentieth century, when the global demand first spiked for Bulgarian yogurt. In short, the global history of yogurt threads through the turn-of-the-century Balkans, to a time when Bulgarian yogurt was at the center of a rising "global" interest in this milky wonder. This history played out in unique ways within Bulgaria itself, where yogurt eventually became a key ingredient in the communist bioimaginary that imagined and then powered a major transformation in the local diet and cuisine.

Fermented milk in its various forms has been consumed for centuries, but the science of probiotics is a twentieth-century phenomenon. In the early twentieth century—not unlike today—the growing demand for yogurt (outside the traditional yogurt belt) was not just about flavor. It tapped into a growing preoccupation with science in general and food science in particular that was—and continues to be—rife with controversies and mythologies. While the language of science has been paramount, it competes with (and sometimes complements) food narratives that look to tradition and authenticity. Such narratives have emerged in the modern era in light of the widespread industrialization and adulteration of food. But their emergence, spread, and place in national and global bioimaginaries are uneven across time and space.

If the Mechnikov moment brought a new food fad to the West, it eventually would provide an enduring scientific foundation for the formation and transformation of the national diet in socialist Bulgaria. For it was only under communism—when Mechnikov's theories on sour milk were pushed aside in the West—that yogurt became a binding agent in Bulgarian cuisine and nutritional science. In part via sour milk, Bulgaria looked to domesticate the ever-changing Bulgarian socialist project, using homegrown models that not only tapped into a "usable" past but depended on it. In a seemingly stark contradiction, a system premised on a "bright future" continually turned to the past for grounding. A society focused on youth looked to the practices of the aged for the key to longevity.

Although the utopian future never materialized, the communist period stabilized and secured a Bulgarian national cuisine—soaked in yogurt—that has proved surprisingly coherent, relatively healthful, and, at least in part, resistant to globalization. Mechnikov's name has reemerged as part of the twenty-first-century return to the notion of probiotics and widespread interest in the human microbiome. But it is only Bulgarians who remember Stamen Grigorov, and they alone take notice of the fact that one of the most important ingredients in Greek yogurt is *Bacillus bulgaricus*.

 CHAPTER 4

"Ripe" Communism

An Ode to the Bulgarian Tomato and Pepper

Contemporary Bulgarian cuisine—with its milky white yogurt and cheese, glistening seasoned meats, and obligatory side of fluffy or flat bread—also looks and tastes decidedly red. This is thanks to the ubiquitous tomato and the long and sweet red pepper, which are the major components of numerous side and main dishes—roasted, stuffed, fried, pickled, or mashed. The bulbous fresh red tomato is now inseparable from Bulgarian cuisine, chopped into the famous *shopska salata* (with cucumbers, feta, onion, and sometimes pepper) and pureed into dips, sauces, and stews. In a 1984 cookbook of Bulgarian national cuisine, tomatoes were included in an astonishing 50 percent of the recipes, with peppers not far behind![1] The smell of roasting red peppers is evocative of late summer in Bulgaria, as is the image of festoons of small hot red peppers hanging to dry. Peppers are consumed in a myriad of ways, including the piquant *liutenitsa*, a delicious vegetable relish with mashed tomatoes and roasted red peppers that (like many Bulgarian dishes) has analogues elsewhere in the Balkans. Bulgarian tomatoes and peppers are legendary both at home and, if only for those in the know, abroad. They are known for their exquisite flavor, a result of Bulgaria's perfect climate for their cultivation—southerly, dry, and hot, adjacent to warm seas, but with sufficiently cold winters. These ingredients thread through and tie together the recipes that today make up Bulgarian cuisine, in every conceivable fresh and cooked combination, as centerpiece, robust accompaniment, or flavor/texture/color infusion.

And yet, it was only relatively recently that the highly nutritious and flavorful pepper and tomato, both of which are New World foods, conquered the Bulgarian (and European) palate and diet. Peppers, especially the small and hot varieties, arrived in the seventeenth century and spread throughout the region. But the tomato and the larger sweet pepper filtered in and took hold much more slowly, gaining momentum in the late nineteenth and the early twentieth century. In the course of those centuries, Bulgaria became known as a cradle for the cultivation, spread, and export of both plants and their products. There was a dramatic reddening of the postwar Bulgarian diet through a veritable explosion in tomato and pepper production, consumption, processing, and exchange. These red vegetables—which botanically are technically fruits—clearly factored into the evolving twentieth-century Bulgarian bioimaginary in important ways. In particular, they provided a dense supply of newly discovered essential vitamins (especially C and D), along with minerals, fiber, and enzymes, to build healthy modern bodies and bolster resistance to deficiency diseases like rickets and scurvy. They were rightfully imagined to supply a source of bodily vitality, necessary to bring progress to Bulgaria as a "backward" nation and, specifically, to build socialism. But they also offered rich and delicious flavor and culinary flexibility to the national cuisine. Tomatoes and peppers thus became an important component of the communist-era Bulgarian diet, which was at once "rational" and deliciously utopian. They became the true king and queen of Bulgarian produce, providing the color and flavor of an imagined bright future.

The reddening of the postwar diet, as part of a general turn toward the increased consumption of produce, was not just a Bulgarian phenomenon. It was global in many respects, germinating and ripening in particular ways in the cultural, political, and climatic conditions of the eastern Balkans and the Eastern Bloc. In most of the world, before the discovery of vitamins in the early twentieth century, fruits and vegetables were largely considered a flavorful luxury, not a sustaining nutritional need. It was only after World War I, and even more so World War II, that a proliferation of research on essential vitamins made fruits and vegetables central to global public health or nutritional thought and practice.[2] Such ideas spread through global science exchanges, but also policy institutions like the League of Nations and the United Nations, which vigorously took up the mantle of optimal nutrition in the shadow of the mass starvation precipitated by the two world wars. Through varied and sometimes circuitous routes, tomatoes and peppers were swept along with the wave of such changes to become prominent culinary undercurrents—in large part via ketchup, salsa, and various tomato and pepper sauces—across large swaths of the globe in the course of the twentieth century.[3] This was propelled and enabled by a combination of changing culinary tastes, new technologies of food processing, and continuing developments in food science. The integration of produce more generally into the modern diet intensified in the Cold War period as dietary science became enmeshed in the biopolitics and bioimaginary of systemic competition. But the new reign of vegetables and fruits was also fueled by generalized postwar prosperity as well as new consumer practices and expectations in both East and West. In short, produce provided variety, color, and flavor to the promised dreamworlds of abundance on both sides of the Iron Curtain.

Bulgarian scientists and policy makers had been part of the global conversation on food and public health since at least the interwar period, a time when growing social concerns mapped onto national ones. If sufficient carbohydrates from bread and protein from animal products were deemed most critical, the nutritional offerings of fruits and vegetables completed the picture, providing essential bio-building blocks and serving as an antidote to diseases like rickets, scurvy, and pellagra. The pepper and the tomato were nutritionally rich, blessed with copious amounts of the essential vitamins C, D, and B3 to hold back the advance of such diseases. Of course, they were by no means the only component of the rapidly filling postwar Bulgarian cornucopia, which also prominently featured cabbage, onion, leeks, garlic, cucumbers, eggplant, and beans. And yet, in all their red splendor, the pepper and tomato became objects of adoration and pride amid this bounty of Bulgarian produce. Their "foreign" origins (that is, the Americas) were gradually forgotten as they became central to Bulgarian national cuisine, as well as key exports—an engine for the postwar economy.

Bulgarians were arguably both beneficiaries of and important agents in the European diffusion of and the growing demand for tomatoes and peppers. With the addition of wine, rivers of red flowed from the Bulgarian headwaters into the Bloc, seeping through the cracks in the Iron Curtain to also redden diets in Western Europe. By the 1960s, Bulgaria was the top source of tomatoes and tomato products in the world, exporting tomato paste, sauce, and puree, ketchup, and canned tomatoes.[4] It also was a major exporter of peppers to the Bloc, while consuming unprecedented amounts of the red vegetables (fresh and preserved) at home. This chapter explores how and why.

Consuming Dialogues

Peppers and tomatoes traveled by boat from the New World to Ottoman shores sometime in the sixteenth and seventeenth centuries, respectively.[5] As noted in past chapters, Istanbul was the mother ship of regional trade, taste, and culinary influence, with its markets' dazzling array of produce, spices, and other foods. The capital's more elaborate Persian- and eventually European-inspired techniques of preparing and processing fruits and vegetables, namely vinegar-based pickling (*turshu*), sugary preserves, and ground spice, became a central culinary model for peoples of the empire. With thousands of ships and land-based caravans arriving each year to this thriving center of trade, foods from the provinces flooded in along with products and plants from the New World. Peppers and tomatoes—along with corn and potatoes, the most successful New World plants—gradually diffused via merchants as well as armies that garrisoned across the Balkans with large quantities of food (and new tastes).[6] Local peasants set up vegetable gardens alongside the Ottoman fortifications that peppered the Balkans, selling fresh produce and other goods to the military. At the same time, specialized market gardens became a common feature of Ottoman cities, catering to developing urban elite tastes. Bulgarian peasants were among the most common seasonal laborers and peddlers of produce in eastern Balkan Ottoman cities.

But it is unclear to what extent fruits and vegetables were part of the diet for most Bulgarians prior to the twentieth century. On the one hand, a range of produce—cabbage, radishes, and a variety of orchard fruits (apples, plums, and apricots)—would have been available in the region.[7] Peasants consumed or sold the seasonal bounty of their own gardens and orchards, which were especially important during the Orthodox meatless fast. Methods of preservation, primarily pickling and drying, also allowed for some off-season consumption.[8] And yet, as described in chapter 1, the scant peasant diet generally consisted mainly of bread, cheese or yogurt, and onion or garlic, supplemented by wine. Most fruits and vegetables were still considered a kind of luxury, if not harmful to the constitution.[9] Onions and garlic were an exception; regarded as medicinal, they were used to battle a variety of ailments.[10] By the nineteenth century, however, commercial revival, social mobility, new seasonal labor patterns, and culinary influences brought an increased variety of produce into local diets, especially for the newly urbanizing Bulgarian merchant elite.

Peppers, in particular, inserted themselves into the Bulgarian diet in a range of varieties and forms, from spicy peppers ground or mashed and sprinkled on bread, to the larger sweet varieties.[11] Hot peppers (*liuti chushki* or *piperki*, literally bitter peppers) were firmly established, if not indispensable, in much of Bulgaria by the mid-nineteenth century.[12] For the Western traveler, it seemed a curiosity that Bulgarian peasants would eat the spicy powder that came from grinding the small red peppers, most commonly sprinkling it on bread.[13] As historians have noted, within early modern European food narratives, hot red peppers had an association with danger—fire, blood, and bodily passions—and represented a kind of affront to the more "subtle" flavors of northern cuisines and cultures. Such perceptions continued well into the twentieth century, if not the twenty-first. The northward culinary journey of hot peppers was slow and uneven, and peppers in more northern climes, as elsewhere, were also eventually "tamed" into less piquant varieties.[14]

Sweet peppers would later eclipse hot varieties in Bulgaria, but not before hot peppers had secured a place in the culinary imaginary as something distinctly Bulgarian.[15] A striking example of this can be found in one of the most widely read and influential literary works in Bulgaria, Aleko Konstantinov's *Bai Ganyo: Incredible Tales of a Modern Bulgarian*. In this 1895 satirical cycle of tales, the main character is a nouveau riche Bulgarian merchant who travels to and through Europe, namely the Habsburg lands. Satire is spun through Ganyo's brutish "Oriental" character, which is enacted, among other ways, through his consuming practices. His lack of decorum and the particulars of his voracious appetite and gastronomic preferences shock the more "civilized" Europeans around him. Peppers enter the story when Ganyo is dining at the elegant home of Konstantin Jireček, a Czech scholar known as the author of one of the first histories of the Bulgarian people (published in 1876) and for his 1881 account of his travels through Bulgaria. Given Jireček's connection to Bulgaria, Ganyo expects the red-carpet treatment when he shows up unannounced at the scholar's home in Prague. Instead Jireček's reaction is lukewarm, even standoffish. Undeterred, Ganyo indecorously invites himself to dinner, where his comical barbarities are put on bold display. Finding the soup far too bland for his taste, for example, Ganyo pulls two dried hot peppers out of his (very Balkan) *disagi* (saddlebags) and crumbles them into the bowl, boasting, "Of course I've got peppers. . . . After all, as you know, 'dear Mother Bulgaria' can't manage without the hot stuff." He encourages his hosts to partake of the spicy supplement, "Bulgarian style!" The hosts decline as the book's narrator offers a wry aside: "Bai Ganyo spiced up the soup so much that a person unaccustomed to it would have been poisoned." Pure fire-eating is on display as Ganyo visibly and profusely sweats while slurping his pepper-charged soup, eventually demanding wine to extinguish the "red-hot iron" in his belly.[16] Ganyo arguably performs the role of the barbaric, hot-blooded southerner with his ingrained desire for heat, with peppers separating Bulgaria from "civilized" Europe.

If peppers (hot or otherwise) did not catch on among the Czechs, they did spread more assertively into Hungary and other parts of Eastern Europe. Most notably, the pepper had inserted itself into Hungarian cuisine by the nineteenth century as paprika (from the Slavic *piperka*), with Bulgarians playing a decisive role.[17] The pepper, in its role as spice more than vegetable, was diffused via Bulgarian itinerant gardeners-*cum*-greengrocers who established small but productive plots in an archipelago of cities in the Ottoman, Russian, and Habsburg Empires. Since the 1760s, Bulgarian migrants had worked for Greek gardening and greengrocery businesses in Istanbul, where they were exposed to a wider assortment of produce.[18] The urban garden model became an important economic niche for Bulgarians, including in Istanbul; there were an estimated thirty-five hundred Bulgarian-run gardens in the Ottoman capital by 1864.[19] Over the next century, more and more Bulgarians became seasonal itinerant migrant workers and/or merchants as part of the *pechalba* (literally "profit") phenomenon in the region, in which rooted subsistence farming was giving way to riskier ventures, cash crops, trade, and urbanization. By the 1880s, some twelve thousand Bulgarian men left home from March until November to grow and sell a widening selection of vegetables and fruits, including peppers and tomatoes, for city markets from Serbia to the Caucasus. They brought the flavors and colors of Mediterranean and New World produce to the surrounding new states and enduring empires. In a sense they served as important vectors of plant diffusion and culinary change, bringing paprika to Hungary and peppers (and eventually tomatoes) to northern markets, and in return taking new culinary techniques and tastes back to their home regions.[20]

Market gardeners contributed to the revitalization of Bulgarian commerce, culture, and, by extension, politics. Almost mythic histories have tied gardeners to the late eighteenth- and early nineteenth-century commercial resurgence, the circulation of people, goods, money, and ideas that propelled the late Ottoman Bulgarian National Revival.[21] Interestingly, the ubiquitous traveling Bulgarian gardener provided a potential cover for revolutionaries on the move. In 1876 the famous Bulgarian poet and revolutionary Khristo Botev crossed the Danube with a band of two hundred rebels disguised as gardeners. It apparently was so common to see a large caravan of Bulgarian gardeners that their passage on the Austrian steamship *Radetzky* and entry through the port of Kozloduy went unnoticed.

Only later, inland, was the band cornered by Ottoman irregular troops. Botev was shot and killed, becoming one of the most revered martyrs and legends of the period. Leaving aside such politically significant events, real gardeners also had a yet to be charted role in the birth of a consumer revolution through the circulation of capital, plant and food varieties, and new tastes.

An array of new vegetables and fruits, including larger sweet peppers (red and green) and tomatoes, slowly began to enter the regional diet in line with evolving Ottoman and European culinary practices.[22] European forms, tastes, and techniques trickled into urban Bulgarian foodways in the course of the nineteenth century, as Bulgarians returned from travel and study abroad. Still, the tastes and flavors of Istanbul remained the primary model for the emerging regional cuisine. The integration of produce into the Bulgarian diet was encouraged by such prominent Istanbul-based revival figures as Petko Slaveikov in his 1870 *Gotvarska kniga*—the first Bulgarian-language cookbook.[23] As noted in previous chapters, this cookbook was not concerned with nutrition, but rather with taste, technique, and the adoption of a high cuisine for Bulgarian speakers. Without a doubt, the book's focus was on meat- and dairy-based dishes and methods, with traditional aromatics, garlic and onions, as the most commonly featured vegetables. The book does, however, feature eight recipes in the cooked "vegetable dish" section, primarily featuring pumpkin (or zucchini), okra, and eggplant; five of those dishes also include meat and/or fish. Peppers appear throughout the book, as spice rather than vegetable, heating up rice pilafs, *chorbi* (soups), and a number of *iakhni* (stews), as well as various stuffed dishes featuring eggplant, grape leaves, and tomatoes. A recipe for stuffed tomatoes—with *kaima* (mincemeat), rice, onion, salt, and ground pepper—is virtually the only tomato dish in Slaveikov's book, a mere cameo.[24] It is worth noting that there are no salads in the book whatsoever, though there are various instructions on how to properly store and preserve vegetables and fruits, including pickling and the preparation of sugar-based preserves. Tomatoes, again, are absent from this section. Peppers make an appearance, albeit only as a flavorful accompaniment to cucumbers. Slaveikov's recipes were undoubtedly more aspirational than reflective of the everyday diet of most Bulgarians—an aspiration, that is, for a richer variety of produce in the diet.

It was indicative of what was to come. After 1878, Bulgarians began to urbanize in ever greater numbers, making the transition, at least in part, from food producers to consumers. But new urbanites also commonly maintained small urban gardens in enclosed courtyards; many kept village

FIGURE 4.1. Women string peppers for drying as more drying peppers hang from the eaves of the house in the background. Photo from 1931. http://www.lostbulgaria.com/?p=5190.

domiciles or family contacts in the village with access to plots of land. Beyond garden outputs, in the urban context there was more produce available at local markets, and from 1896 to 1940, fruit and vegetable consumption increased four-fold.[25] This gradually began to include tomatoes, which made more significant inroads into the Bulgarian diet in the early twentieth century. By the interwar period, vegetables were more prominently featured in cookbooks, used in soups and stews, meatless dishes, and a copious array of salads. Tomatoes now regularly appeared in recipes for tomato soups, stuffed tomatoes, tomato sauce, and salads. Peppers, too, enjoyed increased visibility, not just the smaller, hotter *piperki*, but also the increasingly favored larger-size sweet peppers or *chushki*, served roasted, in soups and salads, and stuffed.[26] Neither type was as ubiquitous as in the postwar period, but nevertheless their color and flavor seemed to be firmly in place as part of a new cuisine that was still in flux.

If this shift was also influenced by European culinary techniques and tastes, there was nothing particularly "European" about it. If anything, a produce-rich diet was more in keeping with the well-developed Ottoman urban culinary tradition, itself a synthesis of Persian, Arabic, and Mediterranean influences, which evolved in the context of extensive trade networks. Indeed, the plethora of new dishes that made their way into local cookbooks constituted Ottoman high cuisine (as opposed to peasant fare)—in the broadest possible sense—in form and substance. Nevertheless, European techniques and terms continued to abound in cookbooks and women's magazines in the interwar years, catering and contributing to the development of urban middle-class techniques and tastes.[27]

Interestingly, Bulgarian peasants (still 80 percent of the population) provided produce for such markets, and yet their own diets did not markedly change. Peasants and the recently urbanized working class continued to survive primarily on bread, along with salt, onions, and legumes. In many respects, vegetables and fruits remained a side or accompaniment to bread—which remained at the center of the diet for the Bulgarian urban and rural poor.[28] Perhaps this explains why, in spite of the above-noted increases, in 1911 fruits and vegetables accounted for only 1.6 percent of total food production in Bulgaria; legumes, by comparison, accounted for 2.3 percent. This does not include wine grapes, which were a full 1.3 percent of the total, even after production dropped steeply in the 1910s due to a major outbreak of phylloxera.[29] Fruit trees were common, especially plum trees, as plum brandy (rakia) saw a steep rise in production, especially after phylloxera wiped out large areas of grape production.[30] In effect, alcohol was one of the most common forms of fruit "preservation" for an increasingly impoverished peasantry and working class. This added to the growing concerns among interwar scientists and intellectuals about the interrelated issues of Bulgarian public health, poverty, and the peasant and worker economy.

Beyond taste, changes in consuming patterns were also directly impacted by global food narratives, which evolved rapidly after World War I. Many of these shifts have been outlined in earlier chapters, but here it is worth noting the important place of fruits and vegetables in the scientific thought and the sometimes intertwined religious and social movements of the period. Advances in food science elevated produce to a new level of importance. The discovery of vitamins in the 1910s, by Polish biochemist

Kazimierz (or Casimir) Funk, spurred new writings and discussions within the scientific community about the importance of vegetables, which in the past had largely been seen as useful more as extra color, flavor, or cheap filler. Produce was now seen as having critical nutritional value, including in fighting scurvy and other deficiency diseases. The mass starvation at the end of World War I (see chapter 1), which unleashed revolutionary political forces, was certainly part of the impetus driving global concerns around food, which were increasingly framed in socioeconomic and political as well as scientific terms. Even after the war was over, the interwar economy remained on a roller coaster of economic crisis, culminating in the devastating 1929 global depression. This propelled a continued pattern of global hunger that had its deepest impact among the rural and urban poor, in Bulgaria, across Europe, and globally. The era of the so-called hidden famine heightened interest in nutrition and disease among impoverished populations, where vitamin and mineral deficiency caused diseases like scurvy, rickets, anemia, and goiter.[31] Combining social with nutritional concerns, studies of wartime and postwar Europe looked at the chronic hunger and nutritional deficits of urban workers and rural peasants, the latter of whom were often forced to sell their foodstuffs because of crushing debt. In the 1920s and 1930s, the Health Commission of the newly formed League of Nations launched a number of studies designed to track global consumption of various foods, but also to determine the basic and optimal diet as an objective for preventing disease and improving public health globally.[32] By this period, vegetables and fruits, along with meat and dairy, were seen as an integral part of what would later be labeled the "nutritional transition." In line with assumptions about inevitable or necessary "development," there was a growing assumption that populations would and should shift from a largely grain-based diet to one in which their intake of fats and proteins, along with produce and sugars, would rise precipitously.[33] If there was a growing consensus that such a "transition" should be made, exactly what that diet would be and how such a change could be achieved were open questions.

To return to a familiar figure from past chapters, Asen Zlatarov—public intellectual, biochemist, and the country's first food scientist—was the most important voice in advocating increased consumption of fruits and vegetables. Zlatarov was more than just a food scientist; he was informed by the new global conversation that centered on the need for coordinated studies and solutions around nutrition and public health. To a large degree, his food science was in line with this general paradigm—that is, he believed that a nutritional transition was needed in Bulgaria. For Zlatarov, however, who had socialist sympathies, the socioeconomic dimensions of both the problem and the solution were paramount. It became clear to him in his studies on Bulgarian nutrition that peasants and workers were malnourished, that their diets were lacking in calories, protein, and vitamins, in stark contrast to the relatively well-fed professional classes of Bulgarian cities. This was a liability for the nation, he argued, as it made workers weak and prone to disease.[34] As noted in chapter 1, Zlatarov's left leanings were made clear when he traveled to the Soviet Union in 1936 for a scientific congress of physiologists, which included the famous Russian scientists Nikolai Vavilov and Ivan Pavlov.[35] He shared a private dinner with the great Vavilov, who described his travels in search of seeds, live roots, and whole plants, amassing a collection of over two hundred thousand live specimens, to be housed in a former palace of Catherine the Great. For Zlatarov, the implications for agriculture were astounding. "Can you imagine what that means for the cultivation and hybridization of plants?" he marveled. For Zlatarov, the Soviet Union was the beacon of a future in which science would be elevated to new heights and could be deployed, along with a redistribution of wealth, in the service of feeding the masses. At the elegant closing ceremony of the conference, Soviet premier Vyacheslav Molotov addressed the scientific priesthood of the era, proclaiming that with their input, "science would take the place of superstitions" and be spread to the masses.[36]

In spite of Zlatarov's awe at what he saw of the Soviet experiment, he had distinct ideas about food science. As a vegetarian who advocated a plant-based diet, he felt that the food science of the day—including Soviet food science—was putting too much emphasis on protein, and therefore meat. His view was that a *sustainable* plant-based dietary system made the most sense for Bulgaria in particular. Given the country's relative poverty, he recommended a diet rich in produce, legumes, and whole grains, with ample dairy and eggs. A focus on vegetables and fruits in the diet, he argued, was smart from both a budgetary perspective and a health standpoint.[37] Zlatarov's praise for produce was grounded in the biochemistry of vitamins and minerals, but it also extended to the digestive properties of fruits and vegetables. Their color and flavor, according to Zlatarov, were important in stimulating digestion—perhaps a nod to Pavlov's research on dogs and salivation. Zlatarov's work extended well beyond the ivory tower. In articles in popular periodicals such as *Vestnik na zhenata* (Magazine for women), for example, he called on readers to eat—and feed their children—as many fresh fruits and vegetables as possible while they were in season, including "tomatoes and salads": "Let's get our children used to eating more vegetables, grapes, raspberries, and strawberries, so that they think of these products not as some kind of luxury, but rather as a necessity for life. And not compotes of them, but as they are in nature, rich in vitamins and life forces, carriers of fiber, which is needed in the gut."[38] Zlatarov praised the hot red pepper as having a particularly important role in nutritional and digestive functions, but he also claimed that too much hot pepper could be "dangerous" for foreigners: "If a Westerner got a bowl of soup with our hot chilies . . . he would be injured!"[39] Zlatarov's recommendations tapped into a combination of mainstream global food science and various vegetarian counternarratives, as well as his knowledge of Bulgarian cultural and social assumptions. He also mapped out aspirations for the future, in which a sensible and sustainable vegetarian diet that included ample tomatoes and peppers would prevail.

Zlatarov's biochemical, socioeconomic, and moral arguments were tightly woven into the writings and activities of Bulgaria's interwar vegetarian movements. Bulgarian Tolstoyans, in particular, regularly cited Zlatarov and his ideas about sustainability, as well as the nutritional importance of vegetables and fruits for personal and public health.[40] Along with the ethical, moral, religious, and socioeconomic arguments of the past, the new science of the vitamin and the other touted properties of produce grounded vegetarian thought in a new way. Vegetarian publications were peppered with produce-rich recipes that espoused the vitamin and mineral content of a plethora of soups, salads, and other dishes. Many of these contained green (or in Bulgarian "blue") and red tomatoes and "sweet peppers."[41] It is worth noting that until the interwar period, "blue" tomatoes were more commonly consumed than red, as the latter were considered rotten.[42] By this time, "blue" tomatoes were still present, but red tomatoes were celebrated and promoted as well. In many cases their nutritional properties were explicitly described, as in an article on vegetables and salads that noted, "when preparing salad in the summer, don't forget red tomatoes, which are the richest carriers of all kinds of vitamins."[43] In the course of the prewar years, fruits and vegetables became part of the conversation—and a range of collective visions—about dietary deficiency and social change.

As with other foods, new processing technologies were utilized for produce in an effort to overcome challenges on the production and supply side. Food scientist Ivan Tomov, for example, who had been one of Zlatarov's students, had noticed that Bulgarian peasants were bringing carts filled with tomatoes and grapes to market only to get low prices for them because of the seasonal glut of available produce. It occurred to him that in winter, when such produce was scarce, it would draw a much higher price, and that better methods of preservation would allow for these vitamin-rich foods to be consumed year-round. With that in mind, he traveled to Italy with a colleague in 1930 and spent forty days in Turin and Parma, observing production and preservation methods for grape and tomato juice and tomato puree.[44] Upon returning to Bulgaria, he spent considerable effort unsuccessfully agitating for the creation of tomato and grape cooperatives that would use such methods. Ultimately his labors would underpin the beginnings of a vacuum-preservation industry throughout Bulgaria in the 1930s.[45] The first fruit and vegetable canning factory was actually established in Bulgaria's second-largest city and center of trade, Plovdiv, in 1899. But development of the industry was sluggish prior to World War I. Even with the changes of that period, there were only forty such factories by 1940, which was a modest number by most standards, but nevertheless indicative of a slow change in producing and consuming practices.[46]

Vegetables and fruits were an important factor in interwar shifts in food preservation as well as scientific, culinary, and biopolitical thought. And yet, by the end of the period, the majority of Bulgarians had roughly the same diet as they had in the past, in some cases even worse, given the intensified poverty of the time. Most could not afford the new bounty of vitamin-rich produce in the Bulgarian diet, or the new urban food culture that embraced variety and flavor. If anything, the parcelization of rural holdings, the shift to growing commodity crops, and the cycles of debt that accompanied vulnerability to global markets—to say nothing of the increased consumption of alcohol—did not bode well for the diet of the rural and urban poor. The interwar period and the war years that followed were still characterized by a simple grain-based diet, and nutritional deficit (by modern standards) for the bulk of the population.

The diet of most Bulgarians became even more austere during World War II, as food production was oriented toward supplying the front and the Axis powers (see chapter 1). During the war and in its wake, the immediate focus was on alleviating hunger, and hence on supplying grains for human and animal consumption to the front and to the populations of the Central Powers. Production of produce continued unabated, of course, and indeed it was an important export to Nazi Germany, where rations of fruits and vegetables remained high.[47] During the war, the biopolitics of food revolved around provisioning in a time of crisis, but it was also about mass mobilization and the management of food production and consumption on a grand scale. As the Red Army occupied Bulgaria and the new communist regime took power, they appropriated this mode of operation; the masses were once again mobilized on a grand scale, but now in the name of peacetime transformation. In the immediate postwar period, this was focused on recovery and building new Bloc alliances under a new regime of centralized planning, agricultural collectivization, and state monopolies over trade and industry. In the years that followed, a legion of managers and food scientists would lead the way in bringing a deluge of fruits and vegetables into the postwar bioimaginary, and ultimately into the bodies of Bulgarians.

Red Fruits, Red Science

As noted in previous chapters, the first item on the postwar food agenda was ensuring a supply of bread, while also increasing consumption of meat, milk, and other animal proteins. But produce was an integral part of state-sponsored postwar food narratives about nutrition and cuisine from the outset. Eventually such plans would ripen into a dramatic change in the scale and substance of vegetable production, consumption, and exchange. The juicy red tomato and the hot or sweet red pepper became key ingredients in postwar Bulgarian foodways. Fresh, cooked, and preserved, in every possible form, they provided the central color and flavors of red Bulgaria.

In no uncertain terms, the postwar pivot toward a produce-rich diet was propelled by food science. From the beginning of the period, food scientists were called on to generate and propagate new notions of nutrition and to educate the populations on proper eating habits. Soviet food science was a clear model, but so too were local "progressive" food scientists like Asen Zlatarov, who had clout and prestige and, more importantly, was a Soviet sympathizer who was politically on the left. Zlatarov had died in 1936, but he left behind a cache of writings along with a legacy of students and research partners. Indeed, his student and associate Ivan Tomov (also a biochemist) was commissioned by the Ministry of Health in 1945–46 to write a pamphlet titled *Something We Should Know about Our Eating Habits*. As outlined in its foreword by a prominent Tolstoyan, Nikola Stanchev, the pamphlet was geared toward educating professional and popular audiences in proper nutrition. Namely, it was designed to provide chefs, cooks, and housewives with information on the importance and basic facts of contemporary food science for health and long life, and more pointedly the "good of the nation."[48] In particular, Tomov articulated the idea of "rational eating," which was very much in line with the borrowed Soviet paradigm of "rational consumption."[49] As Tomov interpreted it, rational eating required knowledge of food science, so that scientific principles could guide individuals as well as the collective national diet. He called on readers to use new kinds of scientific knowledge and frame their eating choices around a proper biochemical balance of nutrients. The focus was on getting *enough* of the requisite foods, though overeating was explicitly critiqued.[50]

Interestingly, Tomov viewed animal proteins as secondary to grains and produce in terms of value and necessity. Echoing Zlatarov's interwar vegetarianism, Tomov noted that meats and sweets were *overconsumed* by elites. They were the "master's food," geared toward satisfying taste, but lacking in vitamins and minerals.[51] Instead, Tomov encouraged consumption of foods rich in vitamins—namely fruits and vegetables—which for him were the key "biocatalysts" in the human diet. In essence, in Tomov's view, produce and whole grains were "more rational" than meat, which was "full of toxins."[52] Tomov, like Zlatarov, questioned the daily nutritional requirements as calculated by Western food science. Instead, carrying on Zlatarov's legacy, Tomov proposed a simple vegetarian diet with plenty of milk and eggs as the most scientifically (and fiscally) rational and healthy. In fact, he looked to the austere wartime diet, which was largely meatless out of necessity, as a healthy model for the nation in peacetime.[53] In many respects, the economic advantages of a meatless diet in socialist Bulgaria were clear: it was cheaper and more efficient in the short and long term to feed the nation with a plant-based diet.

But from a political and ideological standpoint, it posed clear problems. First, the Soviets were meat eaters, who wanted to "catch up" with the West in part via protein parity. But dietary restraint was also in clear competition with generating consent and buy-in among the population using abundant and even delicious food as a "carrot." The question was, would the postwar era be one of austerity—a rejection of a luxurious bourgeois diet, and even the continuation of wartime self-denial—as Tomov proposed? Or would it be a period in which an abundance of foods would be available to all—a utopia of plenty? In the end, the socialist bioimaginary of the future presented a kind of synthesis of the two, but in many respects it was the latter vision that carried the day.

Tomov's pamphlet, written in a moment of ideological flux and uncertainty, proved to be too extreme for the emergent postwar regime. During the period of consolidation of power, the Communist Party needed the scientific knowledge and support of left-leaning scientists and intellectuals like Tomov, and they wholeheartedly embraced "rational consumption" and the curbing of bourgeois excess. At the same time, they needed a nutritional paradigm that would build a newly fortified socialist body, and not just satisfy basic hunger as during the wartime period. Ultimately, they wanted to attract supporters by delivering on promises of abundance to the masses, even if it eclipsed, or competed with, notions of restraint. Tomov's pamphlet was published in five thousand copies in 1948, but it was shut out of bookstores and banned from newspaper ads, severely limiting its distribution and promotion. As Tomov later noted with dismay in his unpublished autobiography, the Ministry of Education conveyed to him that the pamphlet did not "harmonize with the Fatherland Front's ideology of food."[54] Presumably, it was not the promotion of vegetables and fruits in Tomov's work that was the problem, but rather the critique of meat, which was central to the regime's postwar vision for a protein-rich future.[55] In fact, Tomov's short work was never reissued; even in 1955, during the post-Stalinist thaw, it was rejected for publication, based on alleged "factual errors" and its deviation from a "Marxist point of view." The latter was particularly infuriating to Tomov, who noted in his unpublished memoir that his work *was* inherently Marxist in its focus on the "rationalization of food, labor, and energy."[56]

In many respects, however, Tomov's and Zlatarov's approach to food science, which leaned toward the equitable, possible, rational, and sustainable, was influential. In fact, the increased consumption of produce became a central feature of postwar Bulgarian foodways, a new biochemical building block of the postwar socialist body. This, of course, was very much in line with *global* food science. In the United States, for example, fruits and vegetables were one of the four "food groups" that became codified in the immediate postwar years by the United States Department of Agriculture. The United Nations' new Food and Agriculture Organization advocated "food management" on the national and international levels, while making clear that as a source of vitamins, fruits and vegetables were necessary to the "electrochemical activities of the body."[57] In a certain sense, aligning, or more accurately competing, with Western standards on nutritional "development" was more critical than coming up with a socialist alternative. If there was such an alternative, it was to better balance a certain measure of restraint with a utopian vision of abundance for all—not just in Bulgaria, but across the entire Bloc, which also set out to feed its population a greater variety and volume of fruits and vegetables. The bright red of tomatoes and peppers was one of the most important colors in this flavorful new rainbow.

Amid the chaos of postwar transition, new plans were hatched for an immediate increase in the scale of produce production and processing. With the wartime "Law of Mobilization" still in effect, Bulgarian produce—previously exported to Nazi Germany—was redirected, with two hundred tons of tomato puree provided to occupying Soviet soldiers, along with four tons of red peppers (and of course thirty-three tons of vodka and other delicacies).[58] In addition, some two thousand kilograms of tomato puree was requisitioned for export, to trade for needed "machines" with partners within the rapidly coalescing Eastern Bloc. Building on interwar advances, Bulgaria gave immediate attention to ramping up preservation methods, namely canning, to increase export capacity. As early as 1947, a comprehensive plan was in place to set norms for what would become a fully developed food preservation industry. This included the import of needed equipment, as well as the organization of courses on food preservation techniques for local producers and in university and institute settings. The plan called for building new processing plants, renovating old ones, and implementing new hygiene standards. It also attended to things as simple as "making sure there are enough glass jars" to supply the rapidly mushrooming canning operations. The goal was not just to increase local consumption of off-season produce—that is, to overcome seasonality—but also to secure a constant (and growing) supply for export to the Soviet Union and the Bloc. In particular, tomato puree and red pepper pulp or mash were seen as important bargaining chips, something that could be traded for a variety of needed technologies and goods to provide the means to build socialism.[59] Still, in the midst of rebuilding basic infrastructure and creating a whole new political and economic system, there was only so much that could be achieved in these early years.

By 1956, an even more concerted effort was made to increase the quantity and quality of fresh and preserved vegetables and fruits. By then the postwar recovery was largely complete, but Stalin's death in 1953 allowed for more flexibility and unique "paths to socialism" across the Bloc. A general Bloc reorientation toward fulfilling consumer needs was coupled with a slow opening of trade to the West. Todor Zhivkov, the Bulgarian communist leader since 1954, rolled out ambitious plans for development at the 1958 Seventh Party Congress, where a "Great Leap Forward" on the Chinese model was articulated. As part of a set of ambitious production goals across all sectors of the economy, there were unrealistically high quotas for Bulgaria's fourth five-year plan (1958–62). These included a massive increase in the production of fresh produce, but also an improvement in the quality of preserved fruit and vegetables. The immediate goal was to vastly increase the supply of produce of all varieties available to Bulgarians and Bloc citizens, who had ever higher expectations for moving beyond the sacrifices called for during wartime and the period of transition to socialism. It was time for the socialist good life to bear fruit as a way of shoring up systemic legitimacy in the aftermath of Soviet premier Nikita Khrushchev's admission of Stalinist "mistakes." Khrushchev's de-Stalinization had provoked strikes in Bulgaria, East Germany, and Poland from 1953 to 1956, and a revolution in Hungary in 1956, all of which were crushed. In light of this, Khrushchev and Bloc leaders looked for new ways to bolster their legitimacy, which included supporting cultures of consumption. Foodways were a key, if not *the* key, arena for this shift to play out.

In reality, however, implementing the proposed massive increases in production was a monumental task, and the fourth five-year plan sent a wave of panic through the higher-ups in the vegetable and fruit preservation industry. As a commission of specialists involved in the industry explained to the Bulgarian Politburo, the industry was simply unequipped to up production by one hundred thousand tons in the course of the next five-year plan. As they explained, even with postwar improvements in the industry, most of the work in the processing plants was still done by hand, and their limited number of machines could only do so much. Bulgaria was light-years behind the West in this respect, as they frankly reported: "With modern technology, tomato puree can be produced in one hour, but here in Bulgaria it takes four to five hours, with many interruptions, and the resultant product is not bright red, but brown. Sterilization should take five minutes with the new technology, but here it takes two hours. . . . Abroad it takes ten people one hour to peel 5,000 kilos of tomatoes; in Bulgaria it takes 915 people."[60] They went on to detail the massive waste in the industry as a result of the shortage of seasonal labor, which meant that unpicked tomatoes and peppers rotted in the fields; in 1955 alone, some three thousand tons of red peppers went to waste because of rot.[61] It was a picture of inefficiency and problems of labor and field-to-factory coordination.

While skeptical of party quotas, the commission did provide an elaborate strategy for a way forward, from securing "reserve labor" for harvests, to bringing in new technologies, to managing the production process. When the plan was fully hatched, it included importing new machinery from Italy and shifting from glass jars to tin cans to undergird production for the domestic and especially the international market.[62] As in the 1930s, Italy offered a ready model for scaling up the preservation of tomatoes, which had become one of Bulgaria's key food exports in the postwar period. As early as 1957, the Ministry of Light Industry sent a delegation of food preservation specialists to Rome for a congress of the food preservation and packaging industry. This was the first in a series of trips to Italy—most notably to Parma, where the tomato puree industry was centered.[63] Specialists within the Bulgarian canning industry recognized Italy both as "competition" for potential European markets and as the best source of technology transfer for technological advance.

Production targets were boosted for more than just produce, but the increases for tomatoes and peppers were unusually high. A 1955 forecast regarding the preserves industry, for example, foresaw a 42 percent rise in production of tomato puree and a 326 percent rise in preserved red pepper within two years.[64] From this period on, the production of vegetables and fruits—with a special place for peppers and tomatoes—became a major target for state investment. As a result, production and export numbers were on a nearly constant climb, with investments in new technologies and methods—from plant hybridization to preservation and packing—within newly integrated "agro-industrial complexes." Such complexes, developed in the course of the 1970s, were largely unknown elsewhere in the Bloc. They entailed the coordination of up- and downstream agricultural inputs, which dramatically increased production capacity overall. At the same time, highly productive small private plots continued to generate a substantial percentage of produce. Even with continued problems and inefficiencies, Bulgaria by most accounts stood alongside Hungary as one of the few agricultural success stories in the region.[65]

The Bloc remained Bulgaria's primary external market, driving tremendous growth in its fruit and vegetable industry. The circulation of tomatoes and peppers, processed as puree rather than sold fresh, became a mainstay of postwar Bulgarian trade with the Bloc. In essence, Bulgaria, with its more southerly climes and longer growing seasons, became a key supplier of this vitamin C–dense red liquid sunshine to the rest of the Bloc. Unlike the United States and Western Europe, the Bloc had precious little citrus (the best-known source of vitamin C) at its immediate disposal, and it was expensive to import. Bulgaria's unique agricultural capacity was essential to the efficient functioning of the Bloc's alternative food system, which in some respects had its own scientific-nutritional as well as culinary-utopian logic. At the same time, it was quite similar to the Western Cold War nutritional paradigm, especially when it came to produce—that is, certain vitamins (like C) are essential, and the more fruits and vegetables in one's diet, the better. But with the Bloc's far less robust trade capacity and less southerly climate and geography, its food system was limited with regard to the volume and variety of its produce. Direct trade links with the Global South were increasingly established in the last decades of socialism, but trade within the adjacent and integrated Comecon remained more feasible and affordable.

A produce-rich diet remained an important Bloc objective for the remainder of the period. Unlike bread and carbohydrates, and like meat and dairy, fruits and vegetables were consistently slated for increased production, consumption, and exchange. This was based on projected nutritional needs, but also consumer desires. Expectations for variety and abundance were continually on the rise, both because of the urbanization, education, and professionalization of a rapidly growing populace, and as part of the state's obsession with competing with the West in terms of calories, protein, and vitamins. At the level of Bloc economic planning, the Comecon continually charted and analyzed levels of fruit and vegetable production, consumption, and trade capacity—and in a sense set out to both predict and cater to Bloc needs and desires, as well as to shape them. It both directed and responded to the rising demand for more fruits and vegetables in diets across the Bloc, while deeming such increases essential to socialist progress.[66]

Bulgaria was a key player in fortifying the socialist world, but also a harbinger of what was to come. By 1970, Bulgarians consumed 176.6 kilograms of fruits and vegetables per capita per year, the most in the Bloc. Hungary was second at 137.3 kilograms per capita per year, with the USSR trailing at only 88.8 kilograms.[67] In addition to being exemplary consumers, Bulgarians produced more than their own "needs"—by 110 percent, to be precise—but this was just the beginning. In the 1970 fifteen-year plan laid out by Comecon planners, fruit and vegetable production in Bulgaria was slated for a whopping 225.8 percent increase by 1985, as compared to 200 percent across the Bloc as a whole.[68] In order to make such increases a reality, the goal was to "overcome seasonality" through preservation technologies but also production advances, such as the use of hothouses for tomato production. There were also new hybrids and extensive breeding programs for tomatoes and peppers, for which Bulgaria became a recognized trailblazer.[69] Science was brought to bear in developing, for example, disease-resistant varieties of tomatoes and peppers, but flavor was never sacrificed. Indeed, particular Bulgarian tomato cultivars are still closely guarded and maintained.[70] New transportation and preservation technologies meant that produce was no longer sold only at local and regional markets, or produced only by itinerant market gardeners. Now boxes of fresh tomatoes could be put on trucks, trains, and even airplanes, along with jarred and canned forms, for easy export.

State-supported science continued to be the handmaiden to agriculture and trade, even an avenue for the promotion of Bulgarian produce. At an International Conference on Vitamins, held in Sofia in 1960, Bulgarian presenter N. Kolev pontificated on Bulgaria's ideal climatic conditions thanks to its location on the border of the Mediterranean subtropics, which along with its *national* culture of creative gardening made it a Garden of Eden for vitamin-packed produce. According to Kolev, five hundred years of Ottoman rule had produced especially great success in vegetable gardening and wine grape cultivation. During this period, he elaborated, Turks avoided these pursuits, while "nameless" Bulgarian peasants carried out the "selection" of superior seeds and varieties, "perfecting" the "taste, nutrition, and commercial properties" of fruits and vegetables. Kolev highlighted Bulgarian peppers, which by the eighteenth century were spreading throughout neighboring countries, and were known for not just their taste, but also their healthful properties. Beginning in the 1940s, he noted, and going much further during the postwar period, Bulgarian research would reveal that peppers had more vitamin C than virtually any other fruit.[71] Indeed, after 1944, Bulgarian research on the vitamins in fruits and vegetables increased significantly, with both peppers and tomatoes taking center stage. This work had begun to target pepper and tomato varieties for their optimal vitamin C content, along with tracking the best ways to maintain vitamins in picking, storing, and preserving. As Kolev noted, for domestic consumption as well as export, Bulgarian scientists had enjoyed considerable success in maximizing the vitamin C content of peppers and tomatoes, both fresh and preserved.[72]

As Bulgaria embraced its role as produce (and vitamin C) exporter to the Bloc, it also looked to the West for markets out of necessity—as a way to earn much-needed hard currency for further investment in the technologies required to increase production. The international marketing of Bulgarian produce was a regular occurrence during this period, including, for example, in the magazine *Bulgaria Today*. This glossy illustrated quarterly was published in English, German, and French and distributed internationally as a way to propagate a positive view of Bulgaria and its many achievements under socialism. The magazine also was clearly a marketing tool for Bulgarian tourism and the country's exportable products. Articles such as the 1971 "Bulgaria: A Country of Vegetables" detailed Bulgaria's exceptional climatic conditions: "mountain ranges that cross her territory encompass warm fields and valleys, well irrigated and bathed in sunlight."[73] As the author describes, Bulgaria had always been "known for its fruit and vegetables," which were in high demand in the USSR, West Germany, Austria, Switzerland, Norway, Denmark, and England. The article boasted that the state monopoly Bulgarplodexport had exported 280,000 tons of produce in 1971, up from a mere 12,000 in 1939, and that Bulgaria was one of the top producers in the world of field-grown tomatoes and the Danube and Kapiya varieties of peppers, whose vitamin C content was "three times higher than [that of] lemons."[74] Ads for Bulgarplodexport and Bulgarkonserv (the preserved produce monopoly) were also prominent in most issues of *Bulgaria Today*. It was hard to miss the full-color spreads showcasing deep red and pale green bundles of fresh peppers; the large glass jars of tomato paste, tomato sauce, and whole tomatoes; and the pickled or canned mixes of vegetables, such as "Danube salat" (tomatoes, peppers, cabbage, carrots, celery, sugar, salt, oil, vinegar, and parsley). Bulgarkonserv jars were often labeled in English, German, and Bulgarian, clearly ready for immediate export to the West.[75]

For various reasons, however, the American market remained elusive. After resuming diplomatic relations with the United States in 1959, Bulgarian officials began to put out feelers for possible avenues of trade for their country's prized, potentially exportable products. On his first trip to the United States in 1960, Todor Zhivkov traveled around the country looking to promote trade with American businessmen and gain Most Favored Nation (MFN) status for Bulgaria so that it could trade on an equal footing with European competitors. Among other things, he toured a range of fruit and vegetable processing and packaging facilities, where he was reportedly inspired by the possibilities of improving the Bulgarian food-processing industry. As reported in the *New York Times*, Zhivkov announced to his American hosts that he was interested in introducing Bulgarian canned fruits and vegetables—which he touted as "the best in Europe"—to the US market. Bulgarian tomatoes, he noted, were competitive with Italian tomatoes in Europe, but were virtually banned from the United States because of tariffs. Bulgarian imports

FIGURE 4.2. Farmworkers alongside a conveyer belt sort tomatoes for export at a cooperative farm in Bulgaria, circa 1960. Keystone-France, via Getty Images.

into the United States could provide the basis for Bulgarian purchases of US equipment—primarily for packaging and canning.[76] The possibilities of exports to the United States seemed promising, given the results of an East European trip in 1963 by Secretary of Agriculture Orville Freeman and an accompanying delegation of agronomists and economists. Bulgaria was one stop on this fact-finding mission, which focused on agricultural practices and products in the Soviet Bloc and Yugoslavia. In a memo to Secretary of State Dean Rusk at the conclusion of the trip, Freeman stated that "Bulgaria is far behind in agriculture . . . but we saw examples of their best and noted real progress. Some of their agricultural scientists are doing excellent work—notably illustrated by the development of an excellent hybrid tomato, and with progress in other fruits and vegetables." Delegate Dr. Byron Shaw was equally enthusiastic, noting that "Bulgarian tomatoes [were] the best he had ever seen for quality and yield," and that the United States could benefit from Bulgarian knowledge when it came to tomatoes.[77] Unlike Poland, Romania, and other Bloc states, however, Bulgaria did not receive MFN status until 1991, after the collapse of communism, and as a result trade with the United States remained extremely limited until that time.

Bulgaria had greater successes (but also failures) in Europe, as it set out to use produce for barter instead of paying in gold or foreign hard currency (as Eastern Bloc currencies remained inconvertible). Mobil Oil's British representative to the Eastern Bloc, Hugh MacInnes Currie, traveled to Bulgaria on numerous occasions between 1968 and 1976 to discuss potential trade deals. As he later opined in a travel memoir,

> Bulgaria, like most Eastern European countries is badly off for western currency and tries to do barter deals, exchanging agricultural produce for machinery, oil and other commodities they do not have. Once, having negotiated a deal for a large quantity of lubricating oil, when the question of payment arose, I was offered several consignments of tomatoes! Having forcefully declined the suggestion, I was later approached in my hotel at dinner by a Lebanese gentleman who suggested that I should accept this form of payment and he would buy the tomatoes from me! Obviously a "put up" job which I refused to consider. I don't think the Mobil Oil Company would have been amused if I had come back with a shipload of tomatoes instead of a cheque. However having failed in their little ploy, they paid up in real money (US Dollars) and duly got their oil.[78]

Currie's words bring into focus a Western condescension toward Bulgaria as a kind of Eastern Bloc tomato republic, mired in Second World poverty and corruption. If Mobil Oil did not take the tomato bait, Bulgaria found plenty of customers in Western Europe for its tomatoes, peppers, and other produce. In order to capture this market, however, Bulgaria was also compelled to offer the West some of its best produce.[79] As Nikita Khrushchev later ranted in his memoirs,

> They [the Bulgarians] ship them [tomatoes] to us, [and] they turn ripe on the way. The result is garbage. The Bulgarians also export tomatoes to West Germany, but what a difference in the fruit! I'm sure that the West Germans wouldn't buy the ones we get. Why? Because there's competition in that market, while we are compelled to buy from our fraternal Bulgarians. They've got nothing else with which to repay their debt to us. And who suffers? The consumer. The one who has to eat these tomatoes—that's who suffers. The Bulgarians don't eat the tomatoes they send us—you won't find garbage like that in their markets. They eat tomatoes picked one day and sold the next morning. They have wonderful tomatoes right there.[80]

Interestingly, for all of Khrushchev's encouragement of industrialized agriculture, he recognized the problems and deficiencies of large-scale agriculture, long-distance export, and dependence on Western capital.

Much of this he realized while still in power, when he claimed to have encouraged the Bulgarians to avoid the "gigantomania" of massive collective farms and embrace the smaller-scale gardening traditions of truck farming.[81] Such encouragement might have been linked to his own memories of Bulgarian gardeners in what is now Ukraine during his pre-Soviet childhood. Born in 1894, Khrushchev grew up in the Donbas region, which he portrayed as being "supplied with vegetables by Bulgarians," many of them itinerant workers who came seasonally, rented land, and sold their produce on the local market. "They were remarkable gardeners," Khrushchev noted, "the best gardeners in the world."[82] The Bulgarians would sell their fresh produce each morning, he wistfully recalled, calling out to sellers—whom they knew by name—from their horse-drawn carts.[83] Such memories harked back to the Bulgarian market-gardening tradition described earlier in this chapter, which was historically robust in the southern Russian and Ukrainian Black Sea region—so much so that it shaped the future Soviet premier's perception of and even policy toward Bulgarian agriculture. Significantly, in Khrushchev's rant on Bulgarian tomatoes, we find attention to quality over quantity, perhaps fitting for the leader who brought the so-called "consumer turn" to the Bloc at large.

Fulfilling consumer demands continued to be one of the greatest challenges for leaders in the region, especially as they promised that "ripe" communism was fast approaching. At the Twenty-Second Congress of the Communist Party of the Soviet Union in 1961, Khrushchev declared its imminent arrival in 1980. Perhaps the Bulgarian juicy red tomato and zesty red pepper—showered across the Bloc in various forms—were a herald of this "ripe" red future? The Bulgarian system as a whole may have faltered rather than ripened in the 1980s, but Western economic historians declared its agricultural output a relative success story.[84] Bulgaria not only fulfilled its own food needs—it was also able to export 40 percent of its agricultural output.[85] This was possible not just because of the mass production described above, but also because of continued and even expanded small-scale individual production to fulfill local demand. This was also a key element of concerted state campaigns to make tomatoes and peppers an incontrovertible part of the Bulgarian diet, and the national cuisine.

In the Red

In her memoir *Nine Rabbits*, Bulgarian writer Virginia Zaharieva devotes an entire chapter to tomatoes, which seem to spark her nostalgia for the communist past. She details a summer ritual from the 1960s when she accompanied her grandmother to the monastery of the church of Saint Nicholas on the Black Sea coast at "tomato time." As she recalls in vivid detail,

> Around the big table with the vine covered trellis, the nuns were working silently, as if speaking would make the tomatoes go sour. They were making tomato juice; peeled, canned tomatoes cut into large chunks; and thick tomato sauce with celery, parsley, and chili peppers in taller jars. They lugged over tubs of tomatoes; a huge cauldron of water was boiling on the fire so they could steam them for a few minutes right in the tubs to make the skins come off easily. After that, they peeled the tomatoes, cut them into large chunks, and stuffed them into jars. Granny Petrania sealed the lids with a little machine. Her moustache was longer than Grandpa's.[86]

Zaharieva worked alongside the older ladies in this monastic setting, sampling the products of their collective labors. "Sometimes, it seemed like I could hear the red juice pulsing in our veins," she dramatically recalled, and "from time to time, I noisily slurped up the thin streams of juice that trickled all the way down to my elbows, or ate the tomatoes whole when little ones came my way."[87] Tomato time was clearly not just about food; it was tied to social ritual, part of which was consuming tomatoes fresh from the vine and boiled from the pot.

The eating did not stop at the canning table; nor was it limited to tomatoes. But tomatoes were the star ingredient of the "red soup" that the nuns simmered in large cauldrons, which also featured a sweet pepper, carrots, onion, celery, and basil, two nests of vermicelli noodles, a sugar cube, and (oddly) a "cone of incense." A homemade hot pepper sauce was the obligatory accompaniment—"extremely" hot peppers, boiled, chopped, mashed, and strained, blended with brandy, sugar, lemon, salt, and, if desired, basil or cumin. As Zaharieva recalled, "People from the nearby villages who came for the holiday would eat soup, cry from the hot sauce, and put out the fire with the convent bread's thick crust and cold water from the spring—and sometimes with wine, too, if Mother Superior allowed it. The red soup made everyone loquacious and cheerful."[88] Peppers provided a piquant twist to heat up Zaharieva's "tomato time," an early fall ceremony in much of Bulgaria—a practice that connected urban and rural foodways.

In the socialist period most Bulgarians still preserved their own tomatoes, peppers, and other vegetables and fruits. If they did not grow the produce themselves, they bought it at the market or obtained it from friends and family. This "economy of the jar" is often cited as a sign of the systemic failure of socialism to provide the basics.[89] But in fact it was an integral part of the state plan to mobilize producing capacity, as well as an avenue for providing the necessary vitamin-packed consumables. Official publications actively encouraged canning and other preservation techniques.[90] Drawing on tradition (for example, pickling), these sources encouraged the continuity and expanded use of such methods among urbanizing populations. Cookbooks and household manuals like *Kniga za vseki den i vseki dom* (The book for every day and every home) and dedicated publications like *Domashno konservirane* (Home food preservation) widely popularized home canning. *Domashno konservirane* offers a detailed discussion of biochemical processes of preservation—pickling and canning, but also fermentation, drying, and salting. In a clear effort to mobilize home producers against the limitations of seasonal produce, the book promotes "healthful ways to eat all year round," citing the abundance of vitamins in vegetables and fruits.[91] In the flood of recipes that follow, tomatoes and peppers are prominently featured, alone or mixed with other vegetables.[92] In retrospect, the labor involved in such activities may have added to the sense of the hardships of this period in the eyes of many, along with the task of tomato and pepper picking for the youth brigades called on to do seasonal labor. But such sentiments competed with the pleasures of "tomato time" activities as outlined by Zaharieva. Of course, urbanites had their own canning days, when apartment courtyards were alive with the sounds and smells of boiling fruits and vegetables, with rows of canning jars waiting to be filled. Sweet peppers were often roasted first, in towering *chushkopeks* (pepper roasters), a Bulgarian invention that featured an electric or wood-fired heating element in the middle surrounded by a wire form that one could fill with peppers. The *chushkopek* was a common sight in the summer courtyards of socialist apartment complexes.[93]

Such preservation rituals most often involved produce grown on private small plots. In spite of urbanization and collectivization of agriculture, the Bulgarian gardening tradition—remembered so fondly by Khrushchev—continued under socialism. Tomatoes and peppers were grown not just on large agro-farms, but on leased or personal plots (limited to half a hectare), where citizens were incentivized to grow fruits and vegetables for personal use, trade, or sale. The state allowed and even encouraged this, as citizens could productively spend their leisure time contributing to the state's larger production goals. In fact, a large percentage of produce was grown on private family plots, or in the gardens of newly acquired "villas" (vacation

FIGURE 4.3. Tomato products ready for export to the USSR in Pleven, Bulgaria, circa 1970s. http://www.lostbulgaria.com/?p=5190.

cottages). By the 1980s, an estimated 1.6 million private plots throughout the country produced about a third of Bulgarian agricultural production.[94] The reduced workweek for urban workers, with Saturdays off beginning in the 1970s, meant that leisure time expanded and gardening flourished. The government extended loans, waived taxes, and made equipment and seeds readily available; despite the general mechanization of the food system, the DIY approach was highly effective.

These home-produced vegetables and fruits, fresh or preserved, were circulated through friend and kinship networks and were largely preferred over commercially produced varieties. In fact, within Bulgaria, buying canned produce was relatively rare, as canning one's own was the general practice and the ideal. Although urbanization emptied Bulgarian villages of younger

FIGURE 4.4. An advertisement for tomatoes by Bulgarplodexport, the state produce export agency, in *Bulgarian Foreign Trade*, March–April 1971.

populations, the proliferation of vacation villas and cars, better roads, and public transport made regular urban-rural interaction and exchange possible and desirable, even necessary. Despite being a relatively late addition to the Bulgarian diet, tomatoes and certain varieties of peppers now connected newly urbanized generations to the flavor of the village. In some sense, the state's and citizens' reliance on local production saved the local tomato and pepper from the kinds of catastrophic agribusiness transformations that have rendered so much produce almost beyond recognition in the United States. This practice also allowed the state to export more, both fresh and canned, and hence to bring in more revenue for mechanization in this and other areas of production.

Beyond new production practices, recipes and restaurant menus provided a stimulus for increased pepper and tomato consumption. As Bulgarian national cuisine was codified, tomatoes and peppers were omnipresent, whether fresh in salads or canned, baked, roasted, stuffed, and cooked into soups, *giuvechi* (stews), and dips. Tomatoes were also consumed in the form of ketchup, which became common and popular in the 1960s. It was almost inconceivable to not have a splash of red on the Bulgarian socialist (and later post-socialist) plate. For example, a favorite dip for red peppers and tomatoes, called *liutenitsa*, became exceedingly popular. It is not that *liutenitsa* did not exist before socialism; it did. Though it was entirely absent from Slaveikov's 1870 cookbook from Istanbul, it appeared in an 1896 list of words "for a Bulgarian dictionary" by then famous ethnographer Dimitŭr Marinov. *Liutenitsa*, it should be noted, is a word (and dish) from western Bulgaria, where it took on a number of forms, made with red or white onion, eggplant, and peppers—green peppers in the summer and dried hot red peppers in the winter. It is the heat from the peppers, Marinov notes, that gives the dish its name, which comes from *liut*, "spicy."[95] Interestingly, boiled tomatoes and roasted red sweet peppers are notably absent from his recipe. *Liutenitsa* was also absent from important interwar cookbooks like *Nova gotvarska kniga*, published by the women's magazine *Zhenata dnes* (Woman today). It was only under socialism that *liutenitsa* became a ubiquitous national dish, appearing in the 1955 cookbook *Nasha kukhnia* (Our cuisine) as a mash of sweet peppers, tomatoes, oil, and salt.[96] More often than not, hot peppers were replaced with sweet. In the 1978 *Bŭlgarska natsionalna kukhnia* (Bulgarian national cuisine), however, the focus shifts from one standard recipe to a cataloging of regional *liutenitsas*—two from Thrace and one from Stara Zagora. Only one of the Thracian recipes calls for hot peppers, but seven recipes for *liutika* (a pepper mash of varying heat levels) follow.[97] In this period, *liutenitsa* became the most concentrated and beloved manifestation of the complementary sweet and savory tomato and roasted red pepper—a flavor as deeply red as it gets. It became common in its jarred form, as is evident in the 1984 *Gotvarska kniga za mŭzhe* (Cookbook for men), in which three recipes for *liutenitsa* dishes appear—*liutenitsa* with onion, and *liutenitsa* with different kinds of cheeses. All three recipes start with a "jar of *liutenitsa*."[98] Of course, the cookbook was for men, so the expectations were clearly lower—the focus was on ease of preparation.

In many respects, the tomato and the pepper became the most critical building blocks in the invention of a national cuisine. That is not to say that all dishes (vegetable or otherwise) or preparations were pure inventions. Rather, they were gathered in their many local and regional forms and codified into a new *national* cuisine, in which some dishes were wholly new. An important example was *shopska salata*, the Bulgarian summer salad extraordinaire, made from chopped tomatoes and cucumbers (and sometimes peppers and onions) with grated feta on top. As prominent Bulgarian food historian Stefan Detchev has shown, the *shopska* made its first appearance in its current form only in the 1960s, invented and propagated by the state tourist agency Balkanturist for mass consumption by tourists as well as Bulgarians. Its colors (green, white, and red) conveniently echo the Bulgarian tricolor flag.[99] The tomato itself, as we have seen, was not a "traditional" vegetable; nor were fresh salads on the menu until the twentieth century. But under communism, tomatoes and peppers entered the collective imagination as an integral part of the national cuisine, a source of identity, pleasure, and pride.

Food has always been the stuff of memories, an individual and collective conjuring of the past. And for Bulgarians, tomatoes and peppers are among the foods that have a special place in memories, identities, and also nostalgia for socialism. Recall the mouthwatering "tomato time" described in Zaharieva's *Nine Rabbits*. And there is also Kapka Kassabova's lovely memoir, *A Streetcar with No Name*, in which she recalls her aunt and uncle's farm, "where the biggest, juiciest watermelons and tomatoes in Europe grew."[100] As a child she had imagined the Soviet Union as "a very cold Bulgaria on a massive scale, but without the watermelons and tomatoes . . . a place where people ate desperate things like black bread and black caviar."[101] In postsocialist writings, food is a central feature, especially the amazing red flavors of the past.[102] Of course, tomatoes and peppers were also subject to various processes of standardization and industrialization under socialism. But nevertheless, they largely maintained their flavor and texture—unlike, for example, industrialized tomatoes in the United States. This was a Bulgarian point of pride.

Many see the integrity of such local foods as lost, or at least under threat, in the post-socialist era of globalization.[103] After communism collapsed in Eastern Europe, the "economy of jars" did not go away. If anything, it initially gained in importance, given mass inflation and the economic devastation of the 1990s. Since then, the influx of cheap Western food and the collapse of the local food industry have left a void, which, for some, self-production continued to fill. In the post-socialist era, the fruit and vegetable sectors (with the exception of grapes for wine) have fared even worse than other foods. Bulgarian tomato production peaked in 1989, with 874 metric tons produced, only 20 percent of which was consumed domestically. By 1992 exports had fallen to zero, and by 2017 Bulgaria *imported* a whopping 60 percent of all tomatoes and 35 percent of peppers, a figure shocking to most tomato- and pepper-proud Bulgarians. Bulgaria is no longer the headwaters of a red river flowing across the former Bloc. Still, peppers and tomatoes have a permanent place on the Bulgarian plate and in the Bulgarian bio- and culinary imaginary.

Tomatoes and peppers still reign supreme in twenty-first-century Bulgarian cuisine. But this is not a relic of national culture or deeply embedded tradition. Both New World fruits have roots in Bulgarian soil that go back to the seventeenth and nineteenth centuries respectively. And they have flourished as plants and products, drawing particularly good flavor (and nutrients) from Bulgaria's favorable climatic conditions and its gardening tradition. Bulgarians have also played an important role in their European diffusion. But it was only in the twentieth century that they became central to the nation's cuisine. This happened in the context of influential global discoveries within food science, especially of vitamins, which made fruits and vegetables vital chemical elements in the biopolitics of development: fortification for the *modern* human body. This paradigm, which made inroads into interwar Bulgaria, became critical to the postwar regime's ideology of food and nutrition. In order for the country to achieve progress, and even overtake the West, vitamins were as necessary as protein. As with meat and dairy, state actors felt compelled over the course of the period to ramp up produce production and consumption—as bodily fuel for progress, not just for Bulgarians but for the Bloc as a whole. Exports of fruits and vegetables also contributed to Bulgaria's objective to keep moving forward, providing the capital inputs needed to continue to reach ever higher production quotas.

But fruits and vegetables such as tomatoes and peppers were not just for feeding socialist bodies; they were also a flavorful and colorful harbinger of progress, and hence legitimacy, on the Bulgarian plate. If in the past the bulk of the population had had a limited variety of produce in their diet, the socialist period brought a dramatic change, with far more abundance and choice. Expectations changed over the course of the period, as cookbooks and other publications encouraged, and restaurants offered, a more varied diet—replete with vegetable dishes. Through such avenues, new food narratives called for more consumption of produce, not just because of its healthful properties but because of its rich flavor and its purported place in *national* tradition. Tomatoes and peppers, in particular, were infused with meaning, as their rich flavor and color became the stuff of memory. As expectations rose over the course of the period, however, the availability and choice of produce admittedly often faltered, owing to various inefficiencies and problems in production and distribution chains.

This is in stark contrast to the fate of the postwar American tomato. New methods of industrialized farming in the United States have made for a big, beautiful, and machine-pickable tomato, but they have rendered it almost completely flavorless, not to mention nutritionally bankrupt. "Real" tomatoes were exceedingly rare in the Cold War United States, and the vast majority of fresh tomatoes were bred to be picked by machine and ripened with ethanol. They were stripped of flavor and nutrition, unless you were lucky enough to procure some from your grandma's garden.[104] In the United States, the "real" tomato has only been resuscitated as part of the food quest culture of the late twentieth and early twenty-first centuries. It was not that Americans could not grow, or import, real tomatoes and peppers; rather, even in the mid-1990s, these were rare finds on the fringes of the food scene, which only exploded later in that decade and especially in the 2000s. We were—and frankly still are—largely under the dark cloud of the agro-industrial complex. If we take the form and essence of the actual tomato or pepper as a measure of "success" or "failure" during the Cold War, I think we can safely say that the United States failed. If Bulgaria succeeded, however, the ripe tomato and pepper were not enough to replace "ripe" communism.

CHAPTER 5

Wine and Dine

Reds, Whites, and the Pursuit of Bacchus

In 1980, three journalists from the city of Plovdiv set out on a journey to chart the geography and history of Bulgarian wine. They spent nearly a year traversing the villages of the country's far-flung wine-producing regions, from Thrace and Macedonia to the Black Sea coast and the wide Danubian Plain. They interviewed elderly residents, people who, they noted, "had already come to the end of their life's vine and now were tranquilly drinking their wine."[1] The resulting publication, *Kniga za vinoto* (The book of wine), waxes nostalgic about the history of viticulture, winemaking, and drinking in the region. The authors reconstruct this history in vivid and wistful detail, drawing on the living memories of the aged villagers surrounded by ancient vines and barrels full of aging wine. But they also weave in impressionistic vignettes that dig deep into the region's wine-soaked past, back to the ancient Thracians, with one of the earliest known wine cultures.[2] In this late socialist celebration of wine, elderly and ancient makers and drinkers of the bewitching beverage offer an unbroken lineage for Bulgarian wine culture. Far from being jettisoned as retrograde by these socialist-era writers, Bulgaria's wine past evokes longing, pleasure, and most importantly a pedigree of Europeanness.

It was probably no coincidence that *The Book of Wine* and scores of other published works were released in and around 1981, the thirteen-hundredth anniversary of the founding of the first Bulgarian state. This jubilee was celebrated through an astounding number of publications, events, exhibits, parades, and ribbon cuttings for new buildings and monuments.[3] Such commemorations were part and parcel of the heightened nationalist tenor in late socialist Bulgaria, with echoes elsewhere in the Eastern Bloc. Interestingly, scholarship on this period has highlighted the use of nationalism, but also the turn to consumer goods as "alternative sources" of legitimacy in the Bloc beginning in the late 1950s.[4] Yet only works in food studies, it seems, have begun to connect the two parallel processes, most notably in relation to the articulation of national, ethnic, or multinational cuisines.[5] Works on the cultural history of alcohol in the socialist Bloc are still relatively sparse outside the Russian context.[6] But without a doubt, alcohol was integral to leisure consumption across the Bloc, emblematic of the socialist good life, with its promise of abundance, pleasure, and progress for all.[7]

In Bulgaria, as elsewhere in the Bloc, production and consumption of wine and other alcoholic beverages grew exponentially in the 1960s and 1970s. If in the past wine had been simply produced and was one of the core components of Bulgarians' daily diet, under socialism it became standardized, more mechanized, and branded. Production numbers rose steadily, but the biggest change was in the mass export of wine to the Bloc, as well as a focus on quality, with the expansion of new varietals and the use of new techniques and appellation. There was also rapid growth in the production and consumption of a range of spirits—especially rakia (plum or grape brandy). Such beverages were imagined not just as national, but as part of a modern European and civilized lifestyle. In a sense, alcohol offered a pleasurable pairing with socialist progress. It was part and parcel of the intoxicating utopian promise of socialism, or perhaps its replacement. Scores of new drinking venues opened in Bulgaria's urban areas under socialism and at tourist sites, which mainly housed Bloc visitors, mostly notably on the Black Sea coast and at alpine resorts. Wine and spirits were served up to domestic and foreign tourists, but also exported across and on beyond the Bloc in ever greater quantities.

But what of wine's place in the Bulgarian bioimaginary—that is to say, its biochemical effects on the body? Because of its pharmacological effects, alcohol was distinct from other food and drink in this regard. Just as wine culture was bolstered by drinking tradition, many had long viewed alcohol as destructive to the individual body and national well-being. Like wine culture, temperance had a long, even ancient (if intermittent), history in the lands that became modern Bulgaria. In the nineteenth and twentieth centuries, temperance again took root, deeply embedded in global conversations about the psychogenic as well as physical effects of alcohol. As a depressant, alcohol provided an important complement to the other widespread "drugs" of the modern era—coffee, tea, and cigarettes, the stimulants that arguably helped drive the quick pace of modern life, with its emphasis on productivity, work, and speed. Alcohol, in contrast, offered a means to relax or escape the stresses of body and mind, facilitating a needed shift to presumably deserved leisure and pleasure. But was this escape deserved and healthy, or was it a likely gateway to bodily and societal destruction?

Under socialism, drinking became particularly riddled with contradictions. On the one hand, Bulgarian state-sponsored sources encouraged the mass production and consumption of alcohol, and (at least in the later decades of the period) waxed poetic about the history of wine. At the same time, alcohol consumption, instead of being subject to the Orthodox feast and fast, was regulated in new ways. The socialist state sought to limit alcohol consumption to *after* work, to designated leisure times and places. In addition, in the immediate postwar period and again in the 1960s to 1980s, the state supported advocates of temperance against alcohol as a "vestige" of the bourgeois past. This work reached a feverish pitch by the 1970s and 1980s, as a torrent of academic and popular sources, films, talks, and exhibits railed against the physical and moral dangers of alcohol. Voices against drinking were in line with Bloc-wide concerns about alcohol's deleterious effects on the body, the family, the nation, and productivity—the very success or failure of the system. But this coincided with a crescendoing of wine and spirits culture in Bulgaria, which was perhaps more in tune with the general population's love of wine and spirits.

The desires and questions raised by both drinkers and teetotalers before and during state socialism cut close to the heart of Bulgarian history and culture—its past, present, and future. Was wine indicative of a deeper, even ancient, European pedigree for Bulgarians? Was it a critical marker of modern consumer culture and connoisseurship, to be served up to foreign visitors, marketed, and exported? Or was it to blame for backwardness, laziness, wasted time and bodily strength, degeneration, and overconsumption? Was wine a necessary accompaniment at the table of progress and a toast to a bright future? Or was that seemingly bottomless glass of red preventing that future from becoming a reality, or an escape from its apparent failures?

Wine before Communism

In Ivan Vazov's famous novel *Under the Yoke*, set in Ottoman Bulgaria in the months leading up to the 1876 April Uprising, he offers this authorial aside:

With all its hardships, [Ottoman] oppression has this one advantage: it makes a nation merry. Where the arena of political and scientific activity is closely guarded, where desire for rapid enrichment finds no immediate stimulant, and where far-reaching ambition has no scope for development, the community diverts its energies to the trivial and personal aspects of daily life, and seeks material enjoyment. A flask of wine sipped beneath the cool shade of the willows beside a clear babbling brook will make one forget one's bondage; the native *giuvech* [stew], with its purple eggplants, fragrant parsley, and spicy peppers, enjoyed on the grass under a canopy of branches with the blue distant sky peeping through—this constitutes a kingdom. And if there is even a Gypsy piper present, it is the height of earthly bliss.[8]

In Vazov's mesmerizing imagery, wine and food have a special place. They offer a temporary escape from the Ottoman world, in which Bulgarians were politically marginalized. But Vazov also revels in the resultant immersion in life's simpler pleasures. Significantly, Vazov penned the novel in the post-Ottoman period while in political exile from Bulgaria's first post-Ottoman regime. It is plausible that under such circumstances he was also targeting the "oppression" of the post-Ottoman political and economic order, which provoked widespread disillusionment. For all its anti-Ottoman sentiments, *Under the Yoke* arguably captured an undercurrent of yearning for the simplicity of the Ottoman past, when life had centered on the earthly delights of wine and food—if not a call to look to such pleasures to deal with the problems of the present.[9]

Vazov's prose found resonance at a time when European influence and capitalism were rapidly transforming the face of Bulgaria, especially its cities. As noted in chapter 2, Vazov was also the founder of domestic alpinist "tourism," and in addition to fiction, he wrote a number of travelogues about his sojourns to the far corners of the new Bulgarian principality. In his magisterial work *The Great Rila Wilderness*, he beckons Bulgarians to follow in his footsteps to the very "kingdom" that he describes in *Under the Yoke*. With flask in hand, he details his travels through breathtaking natural beauty, among other things reveling in the simplicity of local customs of eating, drinking, and enjoying life. But he also explores and details homegrown traditions of bodily restraint during his pilgrimage to the Rila Monastery and the nearby cave where the renowned monk Ivan Rilski, later canonized as a saint, lived out much of his life in the ninth and tenth centuries. As noted in past chapters, Rilski was known for his extreme piety and fasting, for temperance as well as vegetarianism. But for Vazov, Rilski was a figure to be not so much emulated as revered, someone who fasted on behalf of everyday Bulgarians, who were then free to indulge.

Orthodox Christians navigated the world of earthly pleasures through cycles of feast and fast. Like meat and dairy, wine was at least in theory prohibited on fast days, which constituted almost half the year. Although the holiest monks refrained from drinking, Orthodoxy did not prohibit wine. In fact, monasteries were centers for wine production. In spite of critiques of drunkenness in early Bulgarian Orthodox church writings, namely the *damaskin*, wine was an important source of revenue for the church. The medieval Bulgarian state, so closely aligned with the church, encouraged alcohol production to line its coffers.[10] For Orthodox Christians in the eastern Balkans, wine was regularly, if moderately, consumed, including as part of feast day celebrations. Alcohol was generally not considered sinful; nor was it relegated only to the evening and times of celebration and leisure.[11]

Wine was therefore a part of everyday life, linked to ritual, celebration, and sociability, as well as a source of sustenance. In many places it was consumed for breakfast or during work in the fields—it was considered fortifying, a tonic for strength and vitality.[12] As the famous Bulgarian ethnographer Dimitŭr Marinov noted, "wine, bread, and salt" made up the basic Bulgarian peasant food "troika."[13] Until World War II, predominantly rural Bulgaria was covered with small vineyards for personal consumption; most peasants had small vineyards for their own use.[14] In 1911, despite a steep drop in production the previous year due to the destruction of vineyards caused by phylloxera aphids, wine grapes still accounted for a full 1.3 percent of food production. This may seem like a low percentage, but with other types of produce making up only 1.6 percent of total food production, almost as much wine was being consumed as all other fruits and vegetables put together.[15] In regions where the vine did not flourish, such as western Bulgaria, distilled drinks such as rakia (brandy made from plums or other fruit) and *masika* (an anisette) became common by the fifteenth century, with their popularity spreading in the ensuing centuries to rival (or complement) wine in many places.[16] Within wine-producing regions, grape-based rakia was often made from the *dzhibri*, the pomace that remained after the wine grapes were pressed, and thus rakia production generally accompanied winemaking. Even as late as 1939, however, per capita consumption of spirits (mostly rakia) in Bulgaria was only 1.5 liters a year, compared to 40 liters of wine, which was still considered a dietary staple.[17] Rakia would enjoy growing popularity within twentieth-century Bulgarian drinking culture, but wine had much deeper roots.

Bulgaria had long been part of the traditional Mediterranean wine belt, where viticulture enjoyed suitable soil and climate as well as ample demand. Wine, it seems, may have been produced in the region of Thrace—now divided between Bulgaria, Greece, and Turkey—even before the drink spread to France and Italy. Thracian grape varietals were the probable original source for ancient Greek cultivars.[18] Moreover, Thrace was the home of the famed ancient cult of wine drinking and excess associated with Bacchus (in Greek, Dionysus). But it was also the birthplace of Orpheus, who inspired an ascetic sect that renounced wine and other intoxicants. While this preceded the migrations of Slavs and Bulgars to the region by many centuries, some of the rituals of the followers of Bacchus and Orpheus seem to have survived in local pagan, and later Orthodox Christian, traditions. Under the so-called First Bulgarian Empire of the eleventh and twelfth centuries, a priest named Bogomil founded the famous dualistic Christian heresy that came to be known as Bogomilism. The Bogomils—dubbed "early Protestants" or latter-day followers of Orpheus by some historians and religious scholars—demanded abstinence from drink, meat, and other worldly pleasures. Few remained in Bulgaria beyond the fifteenth century, but their influence spread across Europe to lasting effect.[19] Within the region itself, it was as if wine (and anti-wine) culture had been grafted onto the new inhabitants of the region like a hybrid cultivar. Or perhaps it was just that wine continued to flourish, and with it its discontents continued to rise and fall.

With the Ottoman conquest in the fourteenth century, Islamic law made the sale and consumption of alcohol illegal for the next five centuries, but only for Muslim populations.[20] Christians continued to produce and consume wine and spirits, a practice that starkly differentiated them from the growing number of Muslims in the region—migrants as well as converts—who were generally sober. During much of the Ottoman period, Muslim men congregated in the village or neighborhood café, while Christians gathered in the local tavern. These consuming venues (if not institutions) served as traditional hubs of leisure, but also as social organizations for the men of these two religious communities. It was only in the second half of the nineteenth century that the distinction between café and tavern began to significantly blur—as Christian coffeehouses opened and taverns began featuring coffee, tobacco, and other traditional coffeehouse offerings.[21]

Bulgaria's drinking culture remained robust, but the seeds of a new temperance mind-set entered the country via foreign influences in the nineteenth century. In this period, Western travelers were often judgmental about the drinking habits of Balkan Christians.[22] This was especially true for Anglo-Americans—especially temperance-oriented Protestant missionaries,

FIGURE 5.1. A woman during grape harvest (staged in national costume), circa 1930s. http://www.lostbulgaria.com/?p=1126.

visitors, and temporary inhabitants—who routinely juxtaposed the drunken Orthodox Christian with the sober Muslim. Many laid the blame for overconsumption of alcohol on Orthodoxy itself. Cyrus Hamlin, for example, was a prominent Protestant missionary in the Ottoman field. (He later founded Robert College in Istanbul.) He first explored the Bulgarian lands from his Istanbul base in 1857, accompanied by a British traveler, the Reverend Henry Jones, later the secretary of the Turkish Missions Aid Society. Their reports detailed with dismay local drinking habits, including the claim that all Orthodox priests and monks were "miserable wine drinkers."[23] Like other Protestant missionaries, they were appalled not only that Bulgarian men of the cloth consumed alcohol, but that "the village priest was commonly accorded the privilege of acting as the local liquor dealer."[24] They were particularly concerned about the apparent Orthodox assumption that wine itself was holy, a gift from God. As the Reverend William W. Meriam, a Protestant missionary traveling in the Edirne region in the 1850s, noted, "On my refusing the invitation of one of these [priests] to drink with him, he exclaimed in astonishment, 'What! are you not a Christian?'"[25] Other sources corroborated that Orthodox populations associated drinking wine with a belief in God (or even pagan gods).[26] As a local missionary reported in an 1888 issue of *Missionary News from Bulgaria*, it was difficult to convince Bulgarians that wine was "injurious," as "most people think that wine is a blessing given to them by God."[27]

Such concerns gave purpose to Protestant missionaries, who saw great potential in the Bulgarians and thus made them a primary focus (along with Armenians) of their efforts in the Ottoman lands. As Rufus Anderson, an important administrator and theoretician for the American Board of Missions, confirmed, "Of the various races [of the Balkans], the Bulgarians . . . claim our first attention." The Bulgarians, he added, lived in "a beautiful region, waiting for the taste and intelligence of virtuous industry to make it a paradise."[28] Apparently producing wine did not qualify as productive "industry." As an 1888 article in the Protestant magazine *Missionary News from Bulgaria* explained, "Bulgaria is a country where the vine is extensively cultivated and where, as in some other countries, the most lucrative way in which the fruit of the vine can be used is to make it into wine or brandy which tend to destroy not only the bodies but the souls of those who use them."[29] In American and British Protestant theory and practice, the sobering of Bulgarian bodies and souls was key to nurturing such "virtuous industry" in the region. Only an alcohol-free body, Protestants assumed, could fulfill the promise of intelligence, diligence, cleanliness, health, nonviolence, and the ultimate prize—spiritual purity. By extension, it was drunkenness that kept Balkan Christians from realizing their potential, including securing independence from Ottoman Muslim rule and partaking in European progress.

After their entry into the Ottoman field, American Protestants set up schools across the empire, and especially in the eastern Balkans, with the goal of moral and material uplift for Balkan Christians. Such schools became central hubs for the inculcation of temperance ideas and practice. Temperance writings were also woven through Protestant publications, including the widely read *Zornitsa*, the longest-running Bulgarian-language newspaper.[30] As early as 1848, American Protestants were translating tracts on temperance into Bulgarian, expanding such efforts significantly by the end of the nineteenth century. Missionaries targeted women in particular with their anti-alcohol message, under the assumption that women had the most to lose from the depredations of drunken husbands, but also with the hope that they held sway over the household kitchen, table, and economy, and thus could potentially effect reform within their homes. US-based mass organizations like the Woman's Christian Temperance Union (WCTU) connected their domestic efforts with the extensive work being done around the world by American missionaries. American missionary women founded a local branch of the WCTU in Bulgaria in 1891. WCTU sources noted that they faced distinct challenges, as "Bulgaria is called 'the garden of Europe,' and vineyards are numerous." Moreover, "wine is said 'to be cheaper than water'" there, they lamented, and "brandy is distilled from wine and, being stronger, its result is universal drunkenness."[31] By the 1880s, temperance societies had produced and distributed some one hundred thousand tracts and posters within Bulgaria.[32] They had considerable influence among a vocal minority of Bulgarians who had converted to Protestantism or were simply heavily influenced by their Protestant education. At the same time, their efforts to convert more souls were often thwarted by their insistence on temperance, which became increasingly associated with Protestantism and even leaving the Orthodox faith. According to Dora Davis in a Methodist missionary tract from 1906, "In a country where wine and all intoxicating drinks are universally used . . . to be known as a Protestant . . . implies, without reserve, that the individual neither drinks, chews, nor uses impure language."[33] If in the past Islam had been the main religion of the sober, in the nineteenth century broader efforts at advocating temperance became associated with Protestantism.

This is not to say that homegrown strains of bodily restraint were absent. As noted above, there was the cult of Orpheus, the Bogomils, and Orthodox monks. In addition, a handful of prominent Bulgarian revolutionaries from the mid-nineteenth-century National Revival period were well-known abstainers. Most famously, Vasil Levski, the most revered of all Bulgarian nationalist revolutionaries, rejected the consumption of alcohol.[34] As one biographer noted, "Levsky himself rigorously abstained from both alcohol and tobacco, both of which he considered to be harmful to the individual and society. He was a firm believer in personal as well as organizational discipline."[35] Similar statements about Levski's "purity," while they certainly lean toward hero worship and mythmaking, permeate sources from his contemporaries. Nikola Obretenov, a revolutionary comrade-in-arms, asserts in his memoirs that Levski "would not touch alcoholic drinks of any kind."[36] It is unclear to what extent Levski and others were influenced by Protestant messaging of the time. But more likely it was Levski's own background as an ordained monk and deacon—in the holy tradition of Rilski—and not the Protestant missionary model that fueled his commitment to temperance.

Still, Protestants provided a new Western model of restraint and sobriety, but it was in direct competition with local assumptions about patterns of consumption in the West. For many Bulgarian observers, in fact, Western consuming practices, including those related to alcohol, were patently unrestrained, indicative of the bankruptcy of the very notion of *tsivilizatsiia* (civilization). By the nineteenth century, Bulgarian elites began to regularly use the term "civilization" with ambiguity (as with Vazov, noted above), considerable mockery, or outright hostility. It denoted, for example, foppery, decadence, and the superficial, even hypocritical, nature of both European and Europeanized local elites (Ottoman / Turkish and Bulgarian). In the seminal texts of the Bulgarian nationalist movement, under the heavy influence of socialist thought, critiques of the Ottoman Empire became inseparably entwined with critiques of the West. In the view of well-known revolutionaries like Khristo Botev, for example, late Ottoman "Westernizing" reform— including roads, railroads, and increased trade—had brought Bulgarians only exploitation, as well as debauchery, drunkenness, poverty, and other negative "attributes of European civilization."[37] For Botev, the peasant represented an "unspoiled" exemplar of the *narod* or nation, buffered from the "foreign" influences (the *tsivilizatsiia*) that infected the city. The peasant connection to nature resonated with a critique of *tsivilizatsiia* that evolved in conversation with schools of thought coming out of Western Europe, the United States, and other global "peripheries," most notably Russia.

In spite of this ambivalence, the West continued to have an allure among local elites, who looked to "civilize" the masses through elevated or *modern* modes of consumption. In terms of food and drink, this civilizing impulse undergirded Petko Slaveikov's 1870 *Gotvarska kniga* (Cookbook), which called for a refinement of traditional local wine culture. Indeed, wine and rakia occupied the bulk of the "drink" recipes in *Gotvarska kniga*.[38] Such recipes were not so much a guide to existing Bulgarian wine culture, but rather the promotion of an elevated drinking culture coming from Slaveikov's elite circle in the Ottoman capital. His section on wines runs the gamut from "pure" grape wine to wine made from pears, plums, peaches, various berries, raisins, and elder, as well as various incarnations of "American, French, and English" wines made from apples (also called cider). Replete with technique, Slaveikov's book explains how to spice wine, while cautioning readers to make sure to use sterilized dishes. Building on local practices, Slaveikov sought to inculcate a more cultivated drinking culture to pair with a developed and varied urban cuisine and a refined palate. The audience for his work included newly educated elite urban women, who were charged with managing their families' consuming practices. As such, they were to be central agents in the civilizing process—the creation of a refined domestic culture of eating and moderate drinking for the new Bulgarian urban middle and upper classes.

If women were expected to play a role in creating a more refined drinking culture at home, they were decidedly excluded from the public drinking culture of the tavern or *krŭchma*. Not surprisingly, the tavern itself became a focus of temperance writings and isolated anti-alcohol protests in the late nineteenth and early twentieth centuries. In 1903, for example, a group of women from the village of Gorna Parasel sent a petition to be read aloud on the floor of the Bulgarian parliament, calling for the closing of their village tavern. It was one of the earliest such petitions to target the tavern, which, they charged, was the root of all evil in the village, the place where families' paltry incomes were spent on prolonged nightly debauchery, deepening the community's crushing poverty. In the winter of 1903, the parliament hall was atwitter about this unusual item on the agenda—on the last day, in fact in the last hour, of the legislative session. With patronizing yells of "Bravo," the male parliamentarians expressed a kind of half-teasing support for the beleaguered women of the village, who felt compelled to air their grievances at the highest levels of the Bulgarian government against the scourge of drinking. The sympathetic elected representative from their district, Dragan Tsankov, eventually took the floor, recounting an incident in which a group of distraught women had approached him while he was touring the district. The women were ready and willing to carry the petition on horseback the three hundred or so kilometers to Sofia. Instead, Tsankov had delivered the important package himself, then helped to usher it through session, where it passed! As a result, the offending *krŭchma* was closed—but in its wake a new tavern was opened, which became "the office of the local mayor and even more central to the town's social and administrative life."[39] In response, the women of the village staged a full-blown sit-down strike, a case that again went all the way to the Bulgarian parliament.

Arguably this incident, and other new temperance initiatives, were not just new responses to local drinking tradition; they were also a reaction to change. In the early twentieth century, temperance continued to expand in lockstep with social mobility, changing taste, and streamlined methods for the production and commodification of wine, rakia, and beer. As new forms of public leisure developed, the alcohol market and industry began to take off in Bulgaria, and cheaper, mass-produced spirits became more widely available in cities and villages. Bulgarians returned from studies abroad, bringing new tastes and consumer desires with them, and new migrants came from Central Europe and elsewhere to open cafés, beer halls, breweries, and other alcohol-related businesses. In the interwar period, in particular, the number of *krŭchmi* rapidly expanded, along with *birarii* (beer halls), *mekhani* (wine houses), and other leisure venues where intoxication mixed with food, smoking, and social interaction.[40] A new generation of Bulgarian vintners educated in France brought home new aesthetics and expertise, and wine growing and production cooperatives rapidly expanded.[41]

In the same period, temperance advocacy swelled, driven by such disparate sources as socialist, Protestant, and other religious splinter groups. Such movements were rooted in the prewar years, but they expanded in the aftermath of the moral and socioeconomic crisis of the continuous conflict and deprivation from the 1912–13 Balkan Wars to World War I and its revolutionary aftermath (1914–18). A wide variety of movements and individuals articulated religious, moral, scientific, and socioeconomic reasons in urging Bulgarians to leave the bottle behind. There were purveyors of temperance, for example, within the influential Bulgarian Agrarian National Union, in power from 1918 to 1923 under Alexander Stamboliski. Many within the agrarian movement associated excessive drinking with the "evils" of the parasitical city, and supported referendums to close taverns in the rural areas.[42] In addition, some of the most organized teetotalers belonged to a vocal segment of the Bulgarian Social Democratic Workers Party, the forerunner of the Bulgarian Communist Party. Party members created their own organizations, which became important front groups between the world wars, especially when the BCP was pushed underground after a failed uprising in 1923.[43] An uneasy cooperation emerged between these leftist temperance advocates and Protestants under the umbrella of the so-called National Neutral Temperance Union.[44] In 1938 the movement was permanently banned because of its apparent communist but also Protestant (and hence foreign) leanings.[45] But advocates of temperance emerged on the right as well, in line with eugenicist thought, but also as part of the growing critique on both the right and the left of the Western influences that were penetrating Bulgaria's cities. In spite of the political divisiveness of this period, many intellectuals and politicians were in agreement about the apparent dangers of modernity and capitalism, and for many, drunkenness was one of those ills.[46]

These sentiments also resonated outside of political circles and schools of thought, as alternative spiritual movements (discussed in earlier chapters) looked to a sober way of life for self-fulfillment and social solutions. Temperance was a central tenet of Bulgaria's Tolstoyan movement, which published articles on the subject in its many periodicals.[47] In addition, the Bialo Bratstvo (White Brotherhood) movement, under the leadership of Petŭr Dŭnov, was firmly anti-alcohol. Significantly, Dŭnov had attended American Protestant missionary schools in Bulgaria, an experience that had clearly influenced his powerful alternative vision of the modern world, which called for temperance, vegetarianism, and natural healing via food. The members of the White Brotherhood, like the Tolstoyans, were influenced by Buddhist thought, but also by theosophy, Protestantism, and Bulgaria's Bogomil tradition. For both these groups, the consumption of alcohol and meat was tied to both immorality and toxicity, as well as to the ills of modernity. Such groups were in many respects on the social fringe, subcultures albeit with influential followers.[48] Their bioimaginary looked to salvation of sorts through pure bodies and minds, coming at a time of national (if not global) soul-searching.

World War II cast its shadow over such voices, as the Bulgarian right entered into an alliance with Hitler, and wartime mobilization, along with survival, took precedence in the war years. Many of these movements were quelled in this period, and again under communism. And yet, the postwar communist state seemed to internalize the conflicting local (and global) impulses and dilemmas associated with an existential questioning of the place of alcohol in their nation's future: was the modern man (and woman) to drink or not to drink?

Bulgaria and the Reds

With the Red Army's invasion of Bulgaria in September 1944, the BCP was well positioned to fill the postwar power vacuum that was created as the discredited wartime regime crumbled. The BCP consolidated its power over the political system, economy, and culture in the following years as it set out to domesticate and implement its Soviet-inspired model for social and economic organization. Efforts to mobilize the population for the envisioned transformation were propelled as much through compromise and promises as through force. Feeding the population was an immediate and continuing imperative, initially with bread, then with protein-rich meat and milk products, and finally with vitamin-rich produce. The objective was to fortify the Bulgarian people in order to build socialism, but ultimately also to provide them with abundance and even pleasure—in short, the socialist good life. The question was, how would wine fit into the socialist bioimaginary? Wine, after all, was not a significant source of vitamins, and its healthful antioxidants would not come to global public attention until the 1990s. Yet it remained a pillar of local foodways and social life, both traditional and new.

In contrast to the articulated need to ramp up the production of bread, meat, milk, and produce, Bulgarian communists were divided on the question of intoxicating drinks. Was alcohol an abhorrent remnant of the capitalist past or an integral part of the bright future? Without a clear and comprehensive policy or blueprint, many looked to the teetotalers among the early Bulgarian socialists, but also to the Bolsheviks. The Bolsheviks had tried and failed at prohibition, given that vodka was embedded in the male working-class and peasant cultures that they were attempting to harness.[49] Once Stalin assumed power in the Soviet Union in 1928, he summarily abandoned any such efforts and instead considerably increased alcohol production in his effort to woo Soviet workers with promises of the good life.[50] But Stalin's approach had by no means become policy within the international communist movement coordinated by the Comintern. As noted above, a vocal segment of Bulgarian communists were deeply committed to temperance well after 1928. Still, this was by no means the universally accepted party line. Others in the BCP forged networks and found solace in the bottle. Bulgaria's most prominent communist, and its first postwar leader, Georgi Dimitrov, routinely raised a glass with Stalin at his famed dacha with his circle of most trusted leaders while in interwar and wartime exile in the Soviet Union.[51] Upon returning to Bulgaria in 1946 and assuming power, Dimitrov did not take a firm stand on alcohol, given the larger tasks at hand related to consolidation of power and socialist transformation.

In the short term, a number of anti-alcohol measures were implemented to appease the anti-alcohol lobby in the BCP. This included the resurrection in 1945–48 of secular temperance associations that had been shut down in the 1930s. Re-formed in 1945, the Bulgarian National Temperance Union began organizing local associations, which produced newspapers, lectures, plays, films, and exhibitions.[52] The movement stressed the destructive effects of alcohol on the body, the family, and society, firmly blaming capitalism, but also religious feast days, which were regarded as irrational and degenerate. The state also put in place the "law on *krŭchmi*," which targeted *krŭchmari* (tavern owners) as the beneficiaries of alcohol sales.[53] The law nationalized taverns and limited their number to one for every eight hundred people in settlements with populations of under thirty thousand, and one for every one thousand people in cities of over thirty thousand.[54] But if in theory such a law placated pro-temperance Bulgarians, its enforcement was inconsistent and short-lived. Furthermore, by 1948, temperance-committed cadres were quietly shuffled aside and their associations were disbanded, while newspapers and other publications were blocked and budgets for such work were zeroed out.[55] Their work was now pushed to the back burner in favor of the more pressing tasks of rebuilding the economy and society in the socialist image. Indeed, the widespread unpopularity of their cause made it a liability to the new government. The party could hardly expect the masses to give up their daily wine consumption when so much else was being demanded of them in the effort to build socialism. In the end, Stalin's pro-alcohol policies carried the day.

One of the first orders of business was the 1947 imposition of a monopoly on alcohol manufacture and trade, consolidated (by 1948) under the newly formed state agency Vinprom. This brought most alcohol production and sales under centralized control, with the exception of small-scale production for personal use.[56] The monopoly provided the state with direct revenue from alcohol sales to be used for "building socialism" instead of lining the pockets of capitalist merchants, Orthodox priests, or tavern owners. Vinprom was gradually given the resources it needed to modernize the wine and liquor industry, which became important for both domestic supply and international trade. The first turning point came as early as 1952, when a special government decree called for the accelerated development of viticulture and the wine industry.[57] In part this was to be achieved through the active consolidation of small private vineyards, an endeavor that coincided with the first intensive drive to collectivize Bulgarian agriculture. But this was just one of many initiatives in the dramatic transformation of viticulture and wine production from a small-scale affair that was predominantly for personal or local consumption to large-scale cultivation, winemaking, and export. Wine production became the purview of a number of academic institutes, including the Sofia Institute of Wine and the Plovdiv Higher Institute of the Food and Beverage Industries. The trade journal *Lozarstvo i vinarstvo* (Viticulture and winemaking) was founded in 1951 to chronicle the achievements in local viticulture and wine science, including the expansion and mechanization of grape cultivation and wine production.

In *Lozarstvo i vinarstvo* and a plethora of histories of wine from the communist era, one can find a familiar narrative of industry progress under the tutelage of the communist state. State documents also paint a picture of pre-communist wine production as primitive and backward, with the grapes grown on patchworks of tiny and unproductive parcels of land, and processed in small wine cellars with virtually no scientific study, mechanization, varietal development, or coordinated production.[58] For most of the period, the pre-communist industry's past is described as decidedly pathetic, something to be jettisoned and overcome with new methods and technologies.[59] As such sources explain, since the early 1950s, wine had been produced in ever bigger vineyards and larger and more "hygienic" wineries.

There were in fact a number of factors to drive a wine boom in the 1950s, including postwar recovery, newly forged trade links within the Bloc, and the death of Stalin in 1953. Following the lead of Khrushchev, Bulgaria's dictator, Todor Zhivkov, implemented a range of new policies of cultural and economic reform. This process of de-Stalinization called for a cultural "thaw" and greater attention to satisfying consumer needs, which historians have dubbed the "consumer turn."[60] This new push to provide consumer comforts and pleasures called for abundance and even luxury, but also the need to "manage" desires through "rational" consumption.[61] The reality was that the urbanization and new educational and professional opportunities of the first decade of communism had brought new socialist consumers to the fore, along with new expectations regarding creature comforts. This included a more than occasional glass of wine or rakia. The post-Stalinist state was eager to provide deserved leisure and pleasure for its citizens, even as it directed and controlled the process. In the course of the 1950s, the state was able to transition from feeding the postwar hungry to wining and dining the new upwardly mobile, the beneficiaries of the communist system. Wine provided an elixir for leisure and pleasure at home, and it became fuel for the export engine that would help drive the Bulgarian economy and industrialization more broadly.

By the late 1950s, Bulgaria had ramped up all aspects of alcohol production. An Institute for Viticulture and Enology established in Pleven became the center of research and development for the industry, with ten satellite stations across Bulgaria.[62] Bulgarian production numbers climbed rapidly, as a million new decares (roughly 250,000 acres) of vineyard were planted between 1951 and 1961, with some 156,000 decares more a year planned for the next five years. By 1964, one hundred new wineries with mechanized bottling and labeling were producing some 230 million liters a year, and this was only the beginning. Vinprom factories mushroomed across the country, the largest of which was built in Sofia in 1962–63, with others to follow.[63] Bulgaria became a major exporter of wine within the socialist camp through the state wine export agency, Vinempex; it was used both for barter in and outside the Bloc and to bring in needed hard currency for the larger project of "building socialism."[64] But it was more than just a matter of quantity.

There were also persistent efforts to cater to the increasingly refined consumer tastes in Bulgaria and across the Bloc, which according to wine industry sources had moved beyond the simple and "unpretentious" desires of the past.[65] Scientists at such institutes had broad access to Western publications and hence were attuned to the latest developments in global wine production and aesthetics.[66] While Bulgarians still appreciated local varietals such as Mavrud, Misket, and Gumza, a range of international varietals began to be cultivated and even to replace them. Merlot, Cabernet Sauvignon, Muscat, and Chardonnay, for example, were more likely to appeal to foreign tastes, as they had better name recognition.[67] There was a marked increase in the production of white wines (over the more traditional red), dessert wines, and champagnes, driven by Soviet and Bloc demand. As early as 1964, thirty-four varietals were being grown in Bulgaria, and by the mid-1980s there were twenty-three Bulgarian brands of wine, with distinct names and assigned appellations.[68] If branding was generally sparse under communism, Bulgarian wines (like cigarettes) seemed to have the most elaborate labels, names, and varieties. German, French, and Italian machinery was imported to maintain microbial control and mix large and consistent batches of wine.[69] In the late 1970s, a barter arrangement with PepsiCo allowed the Bulgarian industry to develop even further, as cola concentrate and American wine expertise were traded for Bulgarian wine to be sold on the American and international markets. As a result of the exchange of ideas and technologies, Bulgaria acquired new wine production know-how, such as how to produce a California-style oaked Chardonnay.[70]

Some Bulgarian sources from the period claim that the nation had become the number one exporter of bottled wine in the world by 1971, surpassing even France and Italy. If this was an exaggeration, it was not by much.[71] Bulgaria was fourth in the world in terms of exports of bottled wine for much of the late communist period. Within the Bloc, it was the largest exporter of bottled wine until the early 1980s, when Hungary caught up.[72] In terms of production, Hungary was always number one in the Bloc, but Bulgaria was firmly in second place. Bulgaria was continually slated for increases in production by Comecon planners, as providing wine to the USSR was clearly part of its specialization within the Bloc's integrated economies and its food and drink system. In 1972, Comecon plans projected a 118 percent increase in Bulgarian wine, while also calling for a 543 percent increase in champagne and a 208 percent increase in beer in the same period.[73]

The USSR was by far the biggest importer of Bulgarian wine, but it was also exported to Eastern Europe and in limited amounts to Western Europe and the United States.[74] In part this was a result of the national wine industry's proactive involvement in the burgeoning global industry, with exchanges of information and technologies, as well as tastings at international symposia and competitions. A vigorous tradition of "socialist competition" and exchange, along with systemic competition with the West, underpinned Bloc-sponsored international wine competitions, which began in the 1960s in Budapest, Bucharest, and Bratislava, as well as at Sunny Beach, a new resort on the Bulgarian Black Sea coast.[75] The 1976 competition at Sunny Beach, for example, featured 140 firms from eighteen countries and five continents. Through such events, Bulgarian vintners studied industry tastes and standards and showcased their own technologies and products, which routinely won awards.[76] By 1971, Vinprom had brought home some 539 medals in competitions from Montpellier to Tbilisi, Bratislava, Bucharest, and Ljubljana.[77]

New modes of wine production were accompanied by a newly refined socialist wine culture, which was part and parcel of the larger efforts to modernize or "civilize" the urbanizing peasant masses.[78] How-to encyclopedias such as *Kniga za vseki den i vseki dom* (The book for every day and every home) targeted women as domestic civilizers, with elaborate descriptions of the etiquette of how to set a table, pour wine, or mix a cocktail.[79] If Bulgaria had inherited the Soviet program of "liberating" women from household labor, the images and advice on the pages of these sources seemed to heighten expectations for the creation of new *domestic* utopias.[80] But utopia was by no means confined to domestic spaces. If anything, its public display was even more critical. The state tourist agency, Balkanturist, was in charge of creating public spaces in which food culture would both shape a modern socialist citizenry and provide them with the pleasures of the approaching utopian future.[81] Food and drink offerings and public consumption venues were among its central concerns, as it catered to the droves of Bloc visitors who patronized the shiny new hotels, bars, casinos, cafés, and restaurants dotting the long Bulgarian coastline.

Food and wine were integral not just to local consumption and export, but also to the burgeoning tourist industry under Balkanturist.[82] Nowhere in the Eastern Bloc was tourism more central to the state economy than in Bulgaria, which drew tourists to its lovely Black Sea shores.[83] Mega-resorts sprang up like new cities on the sands of the Black Sea.[84] Hotels, restaurants, cafés, and bars were stocked with Bulgarian wine, rakia, and other inebriating offerings. For Bulgaria, tourism, like wine, was a key bargaining chip in inter-Bloc trade, and hence it was a critical factor in "building socialism."[85] Intoxicating drinks were vital to touristic leisure consumption and sociability. Tourists were invited to see "socialist progress" on display, as well as to contribute to it by spending money on Bulgarian goods and services like food and drink. It was fitting, then, that Bulgaria held its international wine competitions on the Black Sea coast, where it was able to showcase not only its modern resorts and its wine industry, but also its pedigree, namely the nearby ancient and medieval ruins.

Bulgaria did not have many consumer goods that could be traded on the Bloc and the world markets, but its food and drink were widely desirable. Balkanturist marketed the Black Sea coast to foreigners, with wine and culinary pleasures as an important selling point. Emil Markov's English-language travelogue and cookbook, *Bulgarian Temptations: 33 Illustrated Culinary Journeys with Recipes*, produced for Western tourist consumption, offers insights into Bulgarian wine varieties as well as the ins and outs of the local wine industry.[86] Markov's lush descriptions of cuisine and libations are interspersed with glossy color photographs featuring staged tableaus of food and wine. One such tableau depicts a small inlet near Sozopol, on the Black Sea coast, with a picturesque fishing boat bobbing on blue waters in the slightly out-of-focus background. In the foreground a table with a white tablecloth beckons, with two bottles of white wine towering over nine platters of intricately arranged fish and vegetable delicacies, artfully mounded for visual effect. Wine is one of Markov's "temptations," an obligatory accompaniment to food and a constant in his vivid images, descriptions, recommendations, and recipes that claim to unearth local culinary tradition while also showcasing *modern* Bulgarian food culture and connoisseurship.

Markov bestows on Bulgarian wine a kind of imprimatur, outlining its regional features while mandating its status within a bona fide *national* cuisine with an ancient Thracian pedigree.[87] Indeed, in spite of the occasional nod to the modern, it is the past that dominates this ode to Bulgarian food and drink. According to Markov, wine bound and animated the culture of the ancient Thracians, who left their culinary roots and ancient vines in Bulgarian soil: "The merry temperament of the Thracians of old with their lifeloving god Dionysus and the cult of the Miletan colonists for exquisite cooking are typical features of the Sozopolites today. What is more, in this town there has always been more wine than fish—the transparent aromatic Misket, and chiefly the 'fair wine,' the rosy Pamid juice, which pleases lovers of white and red wines. Even when fishermen cook their soup on the shore, they prefer this wine to brandy."[88] Whether the ancient Thracian temperament was passed on to the Slavs is a matter of speculation, or perhaps imagination. Markov's association of the cult of Bacchus (or Dionysus) with Bulgarians of Thrace may seem strange in a source produced under late socialism. But Markov's culinary odyssey documented and lauded the local "tradition" of imbibing, a revelry of rustic flavor, pleasure, and even bodily inebriation.

Markov was not alone in explicitly drawing on the Thracian past for culinary grounding. As noted in previous chapters, this was also the case for bread, cheese, and other foods. But nowhere was it more prominent than in the late socialist flood of works on Bulgaria's wine history, which by the early 1980s were filled with nostalgia for the wine culture of the pre-socialist past. Such writings ran decidedly counter to the main thrust of wine histories of the 1960s and 1970s, which were dominated by a disparaging critique of the irrational practices of producing (and consuming) wine in the pre-communist era, with unambiguous boasting about communist progress in quality and quantity. At the same time, as early as the 1960s, certain aspects of pre-communist wine production were praised, including early wine cooperatives and turn-of-the-century wine science, which was again being conducted at the original Pleven station.[89] In addition, official sources made claims about wine as a source of health and vitality, looking both to modern scientific studies and to thousands of years of "folk wisdom."[90] By the 1970s, the past loomed even larger in evocations of wine's "ancient roots" in Bulgarian soil, with Thracian wine tradition as a common refrain.[91] Wine branding was attuned to this trend, with popular brands such as Trakia (Thrace) marked by an image of Dionysus/Bacchus. Short articles in the English-language journal *Bulgarian Foreign Trade* went into great detail to make these connections for potential international clients:

> "Since when do Bulgarians have their own wine?" foreign experts and business men about to meet experts of the Bulgarian wine industry will ask. And with good reason, it may be added, for traditions count for much in the case of any type of product, they are to a certain extent a guarantee of its quality. Yet when asked this question, we Bulgarians can give no precise answer, for the beginnings of our traditions in this field are lost in the mist of time. History and archeology, however, have sufficient material evidence to show that the ancient Thracians who in Antiquity inhabited our lands were past masters in viticulture and wine-making. Many wonderful legends appeared about their god of wine and merriment Dionysus, known to us from mythology. The gay festival of winegrowers as well as the art of making wines have come down to us from the Thracians.[92]

By 1981, a spate of publications reinforced Bulgaria's ancient food and wine pedigree. There were discussions of past and present Thracian archaeological finds, which included seeds, urns, and images of grapes and wine from the Bronze Age. But wine culture was also described as integral to the years of the so-called Bulgarian National Revival (1762–1878), as well as the 1878–1944 post-Ottoman period.[93] Following this pattern, the 1982 *Book of Wine* (mentioned at the beginning of this chapter) fully celebrated Bulgaria's wine past, rather than deride it as irrational and primitive. The book scarcely mentions the modern wine industry, instead advancing the claim that every Bulgarian's "favorite variety" is *domashno* (homemade) wine.[94] Furthermore, Bulgarian varietals, not foreign transplants, are the most lavishly described. Mavrud, for example, is touted as "the pride of Bulgaria," grown in the black soils of the valleys, red, strong, and traditionally so thick that it could be "carried in a handkerchief"; and then there is the lighter red Pamid, made from grapes grown in the rockier soil of the hillsides, which conveniently ripen fifteen days before the Mavrud grapes.[95] *The Book of Wine* wistfully describes the September wine harvests of the Bulgarian past, as festive as fairs, when caravans of donkey-drawn carts crawled across the fields and up the hills, carrying peasants in brightly colored clothes singing with their "Orphic voices" as the bells jingled on the donkeys' necks. Each detail in these scenarios from the pre-communist past is richly narrated. The grapes were carried back on the donkey carts each evening to the wine cellars, where bare-legged peasants stomped them into mash and liquid and drank some wine from the previous year's harvest. Local craftsmen filled the barrels, tasting the wines, and then sold them to merchants to be offered for sale in regional and even international trade. *The Book of Wine* also celebrates the Orthodox feast days, when wine drinking (or in practice greater indulgence) and *na-zdrave* (cheers!) were allowed.[96] The most important feast days were marked by pilgrimages to the monasteries, each with at least five hundred decares of

FIGURE 5.2. Staged photograph of young people picking grapes: a 1970s advertisement for Bulgarplod, the communist-era fruit monopoly. http://www.lostbulgaria.com/?p=5196.

its own vineyards, at which times the roads were choked with donkey-drawn carts, with revelers drinking along the way.[97] The "march of progress" narrative, so common to sources throughout the communist period, is completely absent from *The Book of Wine*. Instead it valorizes small-scale peasant wine production of the past, embedded in the fabric of Bulgarian rural life, a fabric that had gradually but definitively unraveled in the decades since the communist takeover in 1945. But first and foremost, it celebrates wine—its depth of flavor, texture, color, history, and meaning for Bulgarians. By the early 1980s, wine's place in the communist good life—which now seamlessly merged past, present, and future—appears to have been secured.

Battle with the Bottle

And yet, it was not that simple. For in spite of the intensive efforts to build up the infrastructure for wine production, the accompanying culture of wine drinking was not without problems. In fact, precisely as alcohol production and consumption were increasing owing to official efforts, a 1958 decree called for a strengthening of "the struggle with alcoholism and drunkenness as foreign to the socialist order."[98] As a result, the Bulgarian Communist Party began to sponsor increasingly vigorous *anti*-drinking campaigns, calling on veterans of the old temperance movement to reactivate their publications and programs from the interwar period and early 1940s. Party organizers appointed a national committee as well as regional and municipal temperance groups to carry out a range of new local initiatives and measures, everything from plays, to brochures, to evening events, to exhibitions and contests, to lectures on sobriety.[99] Such initiatives gained more momentum after follow-up decrees in 1968 and 1976, which included restrictive new legislation regarding public drinking and alcohol promotion and sales, and a stated commitment to enforce the existing law on *krŭchmi* from 1947.[100]

A whole anti-drinking (and general temperance) literature—books, pamphlets, articles, and dedicated periodicals—blossomed by the 1970s. One of the key issues that framed these writings was the notion that excessive drinking was a product of the past, specifically capitalism, and as such it had no place in the presumably rapidly approaching communist future.[101] It was not just excessive drinking that was the target. Many of these sources claimed that *any* drinking would lead to drunkenness and excess, and hence was a liability for socialist morality, the individual body, and the future of the body politic.[102] Public establishments became a particular focus of such invective, as temperance publications highlighted failed local attempts to close bars and taverns and replace them with alcohol-free establishments.[103] Numerous lampoons of the process appeared in the pages of the newspaper *Trezvenost* (Temperance). One image from 1975, for example, shows an elderly couple clapping as they watch the closing of a *pivnitsa* (beer hall). In the next frame, however, they are dismayed to see a new sign being hoisted above the storefront that reads "BAR."[104] Even outside of the abstinence literature, a critique of the pervasive culture of drinking and leisure consumption could be found. A 1964 cartoon from the magazine *Turist* (Tourist), for example, depicts a rather puzzled-looking hiker on a mountaintop, surrounded by taverns, bars, and restaurants.[105] In a similar vein, a *Turist* cartoon from 1975 depicts an innocent young girl, with her suitcase in hand, standing on a sandy beach surrounded by bars, while a man lies drunk in front of her on the sand.[106] The message from the official tourist movement (which claimed Ivan Vazov as founder) was clear: tourism should be a healthy and invigorating outdoor pursuit, not an indoor booze fest. This echoed temperance sources, which

FIGURE 5.3. A wine-bottling factory is shown on the cover of *Lozarstvo i vinarstvo* (Viticulture and winemaking), 1965.

saw drinking as a form of "irrational" leisure consumption that weakened workers' bodies and dulled their minds.[107]

In spite of these state-supported efforts, the temperance crusade in Bulgaria was always a difficult one. Its small but passionate cadre of true believers recognized that they were facing an uphill battle. For one thing, they were well aware that Bulgaria's economy was dependent on the profits from alcohol and tourism. This was true under communism more than ever, and the production of alcohol and the building of booze-serving restaurants and bars, especially in the capital city and near resort complexes, continued unabated. The other issue, of course, was that Bulgarians enjoyed drinking. Not only was it embedded in "tradition," but as some temperance sources recognized, it was a product of their new (postwar) urban patterns of leisure consumption.[108] Under socialism, the mass production and unprecedented availability of wine, beer, and spirits, along with the quickening pace and stresses of modern life, enabled a new kind of drinking culture, in which excess was even more pervasive than in the past. State facts and figures glaringly contradicted the prevailing notion that alcohol and excess were products and a "remnant" of the capitalist past.[109] In addition, temperance writers and activists had to face and counter the seductive wine narratives and nostalgic mythologies that looked to this very past for cultural and national grounding and a cultural pedigree.

In response, temperance writings made every effort to create counter-mythologies, by unearthing and touting the temperance "heroes" of the Bulgarian past. At first such figures were almost strictly socialist, primarily BCP members (or supporters) who had been actively involved in temperance prior to 1944.[110] But temperance mythologies also reached back to nineteenth-century revival figures, who either were avowed abstainers (like Vasil Levski and Georgi Bekovski) or at least offered critiques of alcohol in their writings. These symbolic "founding fathers" of the modern Bulgarian state were, after all, revolutionaries and even left-leaning, if not socialist (along with nationalist), and so they could easily be admitted into this "progressive" anti-drinking pantheon.[111] Bulgarian histories of temperance also dug back even further in time, to Krum, a medieval khan of Bulgaria who banned alcohol in the ninth century, and early writings by medieval Bulgarian Orthodox priests and monks.[112] The Bogomils, a more "revolutionary" religious sect, were also a part of this pantheon.[113] Indeed, in the lead-up to the thirteen-hundredth anniversary of the founding of the state, individual Bulgarian scholars were actively tracing Bulgaria's history of temperance, just as others traced its wine history.[114] The Thracians—best known for their Bacchic excess—were notably absent from official anti-drinking sources.

And yet the Thracians were not left out by other voices of restraint and anti-drinking within the Bulgarian establishment. One such voice was Liudmila Zhivkova, the daughter of dictator Todor Zhivkov. Having taken over as the first lady of Bulgaria after her mother's untimely death in 1971, Zhivkova was highly influential in the realm of cultural policy in the 1970s.[115] She was also a driving force behind academic and popular interest in the ancient Thracians, and a close personal friend of Alexander Fol, one of the key Bulgarian scholars of Thracian culture. Significantly, in his own writings Fol argued that Thracian religion was based on not just the cult of Bacchus, but also the "organized coexistence between Orphic (solar) and Dionysiac (lunar) mysteries."[116] It was Orpheus who offered a kind of ancient lineage of bodily restraint that attracted Zhivkova, even becoming an obsession of sorts. She was opposed to the pursuit of increasing consumption as part of Bulgaria's path to the communist future, looking instead to culture, aesthetics, and spirituality.[117] By the mid-1970s, she was known to be a nondrinker (and a vegetarian) who often fasted, practices that reflected her interest in theosophy and Agni Yoga, as well as the Bogomils, the fasting practices of the Orthodox Church, peasant folk medicine, and the Bulgarian White Brotherhood. Zhivkova was far from alone in her spiritual pursuits; she had a kind of "salon" of like-minded Bulgarian intellectuals, including Fol. In many respects, she was enormously influential in the development of "alternative" strands of culture and science. The imprint of her ideas could be seen in select temperance writings that echoed her rhetoric on aesthetics.[118] Zhivkova was certainly a major figure in the flowering of public interest in Bulgaria's past, as well as the grandiose celebration of the thirteen-hundredth anniversary of the Bulgarian state. But she also presented a clear threat to the communist establishment, with her open spirituality and rejection of the official approach to creating a communist good life grounded in pleasures and intoxicants. Even today there is serious speculation that Zhivkova's mysterious death in 1981 was due to foul play from within the party establishment, who saw her alternative vision as problematic.[119]

After Zhivkova's death, the Bulgarian state's contradictory approach to alcohol continued to intensify. More publications on the glories of Bulgarian wine were released, even as a less than quiet war on drinking was being waged. This war intensified most precipitously when Mikhail Gorbachev assumed the helm in the Soviet Union; decreased Soviet demand had a reportedly dire effect on Bulgarian wine production as large tracts of grapes were uprooted or not replanted.[120] Exports had grown from 200 million liters a year in the early 1970s to over 300 million in the 1980s, but they tumbled to 205 million by 1986.[121] Indeed, if one examines a graph of Bulgarian wine production and export numbers, there is a steep drop-off that starts in 1986 and continues through and beyond 1989; exports fell to 179 million liters in 1989 and hit rock bottom at 57 million liters in 1991, while production dropped to 266 million liters in 1989, dipping further to 168 million liters in 1993. Interestingly, in contrast to meat, milk, and vegetables, these dramatic drops in production and export predate the collapse of communism—and may have even contributed to it. As in other sectors of the economy, the industry was initially devastated by privatization and international competition, but it has recovered to some extent, and wine quality has greatly improved. Nevertheless, production and export numbers remain much lower than under state socialism: 130.5 million liters in 2020 (as opposed to 208 million in 1961), and 30.4 million liters (as opposed to 70.3 in 1961) respectively.[122] As in other sectors of light industry, the shift to capitalism was devastating to an industry built by the socialist state.

The story of Bulgarian wine has always been subject to the global ebbs and flows of ideas, passions, people, taste, technologies, and even plants. Wine was a homegrown source of pleasure, but also a paradox. Questions and debates surrounding drinking predated communism, but they became most pressing in the period when the most dramatic transformations in wine production and drinking culture took place under the specific constraints of a communist regime and the Cold War. The Bulgarian cult of wine was by no means just for tourist consumption; it offered locals a pharmacological escape from the ills or failed promises of the present, just as alcohol offered escape from the pressures and problems of capitalism.

Amid the competing discourses of the 1960s through the 1980s, wine was a critical trope in multifaceted and contested visions of the communist future. These visions were changing, not just because of regime dictates, but because of the desires of a population that had been dramatically transformed by the postwar communist system. Throughout the period, narratives on drinking practices were a critical part of the Bulgarian response and the country's positioning in relation to progress and modernity. By late socialism, the place of wine had been firmly established, sanctioned and encouraged by official production quotas, consumption venues, and a variety of publications. It was interwoven with new kinds of leisure practices, everyday as well as touristic. Bulgarian wine culture, enabled by the communist state, was about fulfilling the everyday expectations of a new generation of urban consumers. For Bulgarians, wine allowed for tradition to be mined or reinvented so as to reinforce or redirect communism's utopian promise.

Wine offered a fulfillment of, or perhaps an alternative to, the failed expectations of a utopian future.

And yet, even as modern communist-era wine culture was bolstered by an ingrained local drinking culture, it was not without its discontents. The notion of pursuing "bourgeois" pleasures under communism was always theoretically and practically circumscribed by official narratives on consumer and bodily restraint. As official sources cast aspersions on decadent "vestiges of the bourgeois past," they directed much of their criticism to wine and other alcoholic beverages. The Bulgarian Communist Party had a strong subculture of temperance, with vocal advocates of abstinence from alcohol who had been active since the early twentieth century. These efforts continued to strengthen, with considerable energy invested in Bloc-wide temperance campaigns by the 1970s and 1980s, precisely coinciding with Bulgaria's crescendoing wine culture. Temperance advocates, too, felt the need to reinvent a long, even ancient, history in the lands that were now modern Bulgaria. Ironically, the closer the presumed "bright future" of communism came, the deeper into the past such conflicting narratives reached, and the more intractable the conflict became. But in the end it was the drinkers who prevailed. Postwar communism had created the conditions of modern urban life, in which drinking had secured for itself an inexorable place.

Conclusion

As Felipe Fernández-Armesto aptly points out in his *Near a Thousand Tables*, food has always been "what matters most to most people for most of the time."[1] In the past few decades, a rich body of food studies scholarship has recognized this fact, teasing out the multidimensional ways that humans have transformed our complex foodscapes, and the ways that food systems have altered our bodies and our environment. Through agricultural, industrial, nutritional, and culinary choices related to food, we are—or become—what we eat, biochemically and culturally. Our decisions also transform food itself, its biochemical makeup, size, shape, flavor, and how it is combined and served. Our foodways have a lot to say about who we are as individuals and as collectives, as unstable as that might be. It is precisely this instability, the transforming and transformative nature of food, that this book explores in the context of modern Bulgaria.

In particular I outline the ways in which the Bulgarian diet, cuisine, and food itself significantly transformed, especially during the postwar period. In order to do this, I bring into relief the distinct histories of Bulgarian bread, meat, yogurt, tomatoes and peppers, and wine, tracing the ways most of these ingredients became central to the Bulgarian diet *only* under communism. That is to say, the communist regime put in place conditions of production and cultures of consumption that ensured that meat, dairy, and certain kinds of produce would be consumed regularly, even daily, irrespective of season or social class. Bread and wine were already a steady part of the Bulgarian diet, but both were fundamentally transformed along with meat, milk, and produce during that same period. If dietary and culinary transformations were under way since the nineteenth century, with the advent of communist rule they rapidly quickened under the aegis of a centralized state that had the resources and the will to fundamentally reshape the way people ate.

Such changes, as I argue, unfolded in the shadow of an ever-changing Bulgarian bioimaginary—that is, ideas about food and bodies shaped by religion, science, or contested narratives about "civilization," tradition, modernity, and development. The bioimaginary also evolved in the context of momentous political and geopolitical transformations, which framed or privileged certain ways of producing and consuming food. Bulgaria was not alone in undergoing sweeping changes to its food system in the twentieth century. Similar changes happened globally, but the pace and path of such changes were ever determined by unique local, national, and regional circumstances. In the Bulgarian case, the articulation of national cuisine happened in lockstep with a "nutritional transition" implemented by a modernizing authoritarian state. Bulgaria provides a fascinating story (a narrative center) in its own right, but also a case study for how global food narratives, policies, and practices impact (and interact with) the "periphery."

In that regard, one of the preoccupations of this book is the twentieth-century transition from religion to science as the predominant "belief" system for the regulation of bodies. Prior to the communist era, religious thought and practice played the dominant role in who ate (and drank) what and when. Across the region, consuming practices divided Muslims and Jews from Christians, with, for example, proscriptions against pork (for Muslims and Jews) and alcohol (for Muslims). My focus, however, is on Orthodox Christian populations whose traditional dietary practices were circumscribed by cycles of feast and fast, which regulated meat, dairy, and alcohol during 182 *postni* or fast days a year. In practice such strictures were not always followed, and seasonality, economics, and a variety of other factors likely had as much of an effect as religion on what people did and did not eat. Nevertheless, religion was a critical factor in framing the bioimaginary that tied the consumption of certain foods and drink (like meat and alcohol) to bodily pleasure and hence sin. Total temperance or vegetarianism was rare, and as such was connected to the heightened holiness of select monks—such as the hermit saint Ivan Rilski. By the twentieth century a number of religious or political movements looked to temperance and/or vegetarianism as sources of spiritual redemption or social change. Some looked to the medieval

Bogomils or the Thracian cult of Orpheus as examples of bodily restraint that had sprung forth from what was now Bulgarian soil. At the same time, by the interwar period, notions of spiritual and bodily purity were increasingly attuned to new global ethical, nutritional, and even environmental understandings of the healthful properties of a plant-based (or alcohol-free) diet. For many Bulgarian vegetarians and temperance advocates, spiritual concerns were fortified by new knowledge about the biochemistry of food and drink, as well as novel frameworks for understanding the politics and socioeconomics of consumption.

In the postwar period, the Communist Party attempted to jettison religious "superstitions" altogether, in favor of food science, which became a handmaiden to socioeconomic transformation. Bulgarian socialists (like the Bolsheviks) were politically buoyed by the mass hunger and lack of bread during World War I and its aftermath. But upon their gaining power in the postwar period, the goal rather quickly shifted to moving beyond bread, turning to nutritional science in an attempt to supplant religion as the organizing principle for a new bioimaginary and food system. The explicit goal was to move beyond the "irrational" patterns of the religious feast and fast, and also to overcome the problems of seasonality and distance. The state sought to feed the entire population in line with the Soviet-inspired notion of "rational consumption," underpinned by the new calculus of food science. Radical changes in production were needed and were made possible by the unprecedented level of state control over agriculture, industry, trade, and academic and research institutions. Coupled with rapidly, and eventually massively, increased production through collectivization and mechanization, food was subsidized as never before, and amply feeding the population became a central part of the communist-era social contract. Such objectives were overtly connected with Cold War systemic competition—to reach parity with the West, or even to better provide populations with a balanced and nutritionally complete diet. Western, Soviet, and Bulgarian food scientists provided a rather similar general formula for postwar nutritional standards. Namely, they favored a protein- and vitamin-rich diet, calling for massive increases in consumption of meat, dairy, and vegetables.

These were the ingredients of change, necessary to fortify Bulgarian bodies, which needed daily fuel to work—and thus to build socialism. These ingredients, however, had to be altered or shifted in order to increase production and fulfill the demands for domestic consumption. In most cases, the quest for quantity dictated shifts in what was produced (and hence consumed), for example a move from lamb to pork, from sheep's to cow's milk, from small and hot peppers to large and sweet varieties. Such changes were by no means aimed only at domestic consumption. Bulgaria was a key producer for the Bloc, especially the Soviet Union, to which it shipped massive amounts of wine, processed pork, tomatoes, and peppers. Bulgaria also provisioned visitors to the country, especially the legions of Bloc tourists who frequented the Black Sea shores and mountain resorts. Here Bulgaria became a kind of showcase for the socialist good life, complete with provisioned restaurants that showcased the new nutritionally balanced (and delicious) national cuisine.

For simply providing a rational supply of essential vitamins or protein was never the end goal. Food was central to the utopian promise of a brighter future that required constant assurances and tangible evidence of progress. If bourgeois excess was vilified, a restrained embourgeoisement for the Bulgarian peasant and working classes was the order of the day. This served both to buy off the masses with material comforts as proof of the vitality of the system, and to "civilize" the great unwashed, who could now embody socialist progress. In large part, women were responsible for this civilizing mission, which included arranging, provisioning, and managing domestic spaces and bodies. Bulgarian cookbooks and how-to manuals provided the basics of nutrition, along with thousands of recipes, menus, and techniques, everything from advice on table arrangements, outfitting kitchens, and rules of hospitality. Recipes and menus featured large amounts of the new ingredients of change: deliciously prepared meats, dairy, and vegetables. Breads and pastries were also featured, now as a pleasurable accompaniment to the diet and not its primary source of nutrition. And then there was wine—of negligible nutritional value, but a critical source of flavor, pleasure, and social release. Gastronomic pleasures that had once been holiday fare were integrated into the everyday, as part of a postwar culture of food that extended to—and in large part developed within—the plethora of restaurants springing up in cities and resort areas. This new food culture held the promise of making Bulgarians *modern*, but the new cuisine was also increasingly framed as *national*. As Bulgaria turned to nationalism as a source of legitimacy in the later decades of communism, a newly minted national cuisine was codified and elaborated. Food and wine narratives now looked to local culinary "traditions" and food history going back to Thracian times (and even Bacchus), to offer a kind of European food and wine pedigree for the Bulgarian nation.

The open embrace of the pleasures of food and drink, flavor and experience along with nutrition, were important to Bulgarian communist legitimacy. But such an approach was never without controversy. Interwar leftist vegetarians and temperance advocates, initially brought into the party's efforts, were pushed aside as early as 1947 in favor of a more meat- and alcohol-heavy approach to wining and dining the masses for popular support. By the 1960s and 1970s, however, new concerns about alcohol consumption translated into far-reaching state-directed temperance efforts. If drinking in moderation was part of the socialist promise, drunken excess was seen as a threat to individual bodies and the body politic. Even as the state supported continued increases in alcohol production and even connoisseurship, it also supported campaigns to curtail drinking and even eliminate it altogether. Temperance messages were spread through books, newspapers, and events such as talks and exhibitions. Meat consumption also began to be questioned, although not in such an organized fashion. As far as the state was concerned, meat production and consumption—a veritable meatopia—continued to be an important and necessary socialist achievement. And yet vegetarian sentiments also came into the public discourse under late socialism. The Bulgarian communist leader Todor Zhivkov's own daughter Liudmila was an avowed (and non-drinking) vegetarian during her years of state influence in the 1970s. During the last two decades of socialism there was a smattering of vegetarian cookbooks and other writings on vegetarianism. Like those who divined a new food culture, temperance advocates and vegetarians looked to the past for precedents. They invoked Bulgarian (and regional) forebearers, from the Bogomils and the cult of Orpheus to interwar vegetarian advocates, to reinforce the *national* scaffolding of Marxist-attenuated ethics of restraint.

But such voices of self-discipline were always in conflict with the state's herculean efforts to fulfill the demands and desires of its increasingly urbanized and educated consuming populace. If voices of restraint (including Zhivkova) suggested a different path to the future, Bulgarian state rhetoric of competition, abundance, "progress," and national greatness set the country on a collision course with systemic failure. From the very beginning of the period, the objective had been to properly fortify the population, to show in facts and figures that Bulgarians consumed as much protein and vitamins—and especially meat—as the most developed countries in the world. It was imperative for the socialist state and its Bloc partners, as well as the food system as a whole, to be capable of feeding their populations not just the basics, but the optimal diet, as in the West. Proving that they could provide all with tables laden with cheap, state-subsidized food and drink was fundamental to the socialist dream, something that set the Bloc apart from the West. Food was progress—it provided the required fuel to build the future and the nutritional and culinary proof that it had arrived. But the state struggled to continually provide the populace with such a diet, which by the 1980s included the most meat per capita consumption in Bulgarian history. In fact, Bulgarian (and Bloc) planners made a serious miscalculation in presuming that they could curb the overconsumption of breads and carbohydrates in favor of more meat and protein-rich dairy simply by cutting or redirecting grain production. Comecon-dictated shifts from grains for human consumption to grains for animal fodder ultimately necessitated massive imports of grain, resulting in ballooning debt across the Bloc. Increased exports, poor planning, and a range of other problems caused intermittent food shortages, which became ever more persistent in the last decades of socialism. Under late communism, the gaps in state supply in Bulgaria were filled—and in some cases caused—by hoarding, filching, and above all DIY production of food on private plots and in private kitchens. Empty shelves did not always mean empty tables, as food was obtained through family and personal networks, or from bartering with known vendors. While such shortages were real, they were relative: it all depended on when and where you were in Bulgaria and what you were comparing them to. To the prewar period, when the bulk of the population lived primarily on bread, wine, and onion, with smaller amounts of dairy, meat, and produce? To the immediate postcommunist period, when local food industries plummeted in productivity as processed imports flooded in? To the 2020s, when shortages are a thing of the past and a more vibrant food scene exists, but when at least some people have to worry about putting food on the table at all?

The communist period witnessed a veritable revolution in food, but the post-1989 period had wrought change that was just as far-reaching. I have witnessed the transformation in Bulgarian foodways over the past thirty-plus years, albeit in snapshots of experience with each of my almost annual visits since 1994. For most of this period, I wasn't studying Bulgarian food per se, but rather enjoying it! As I mentioned in the introduction, Bulgarian food was one of the appeals that drew me deeper into studying the region. I was delighted by its tang, zest, flavor, and texture, but also intrigued by how it echoed other cuisines throughout the Balkans. I watched as the food scene slowly morphed from largely small-scale Bulgarian restaurants and shops with local foods to global products, fast-food chains, and a wide array of ethnic restaurants. Upscale, downscale, kiosks and other forms of public offerings have mushroomed—with some making it and others folding as they rode out the waves of economic crisis that have hit Bulgaria since the 1990s. A seemingly massive number of vineyards and wineries have sprouted up or changed hands, and along with them, a new wine tourism has grown. But still, production numbers are well below the level of the

1980s, and even lower than the 1960s. Local production of what were once Bulgaria's ingredients of change has plummeted in the face of new regulations and cheap imports.

But so too have "slow" and organic food movements emerged, along with a myriad of ways to recognize, protect, and celebrate local foods and food heritage. Over the past decade I have stayed at the agritourist "Wild Farm" in the rugged Eastern Rhodope Mountains, where they raise heritage cows and serve up some of the best sheep's yogurt I have tasted. I have visited the yogurt museum at the birthplace of Stamen Grigorov, among the bucolic mountains of western Bulgaria. Although I did not make it there, Bulgaria also hosts wine, bean, and honey museums as well. I did, however, brave the bumpy, windy roads of the Strandzha—a mountainous region on the southern Bulgarian Black Sea coast—to spend an afternoon reading documents in the Bulgarian Tolstoyan museum. I ate and drank my way across Bulgaria, tasting the changing cornucopia of the present as I studied the past. As the postcommunist food revolution unfolds, Bulgarians too are looking to preserve the ingredients of the past even as they formulate their recipe for the future.

Notes

Introduction

1. I use the term "Balkans" here interchangeably with "Southeastern Europe." I recognize that in some periods in history the term has been employed in a derogatory way, but I use it here in recognition of the local embrace of the term, particularly in the eastern Balkans. For more on this see Maria Todorova's classic *Imagining the Balkans*. For a more recent discussion of the complex meanings of the term see Mishkova, *Beyond Balkanism*.

2. See Lucy Long's introduction to her *Culinary Tourism*, 1–19. It is important to note the association of this impulse with "colonial" impulses to dominate, define, or "consume" other cultures. There is a rich secondary literature analyzing the "Westerner" traveling east, though food and drink are rarely looked at in depth. For a notable exception see the sections on food and drink in nineteenth-century Southeastern Europe in Jezernik, *Wild Europe*, 47–55, 147–70.

3. For a lovely book on this shared food culture that includes recipes see Kaneva-Johnson, *Melting Pot*.

4. These are the Bulgarian, Bosnian/Croatian/Serbian, and Turkish terms, respectively, but a version of the dish can be found all over the Balkans, as well as in the former Ottoman territories of the Levant and Anatolia. For an interesting work that ruminates on the meanings of this dish in the former Yugoslavia see Mlekuž, *Burek*.

5. See, for example, Cwiertka, *Modern Japanese Cuisine*, and Brulotte and Di Giovine, *Edible Identities*.

6. Some of the best interdisciplinary introductions to the field are Belasco, *Food*, and Albala, *Handbook of Food Studies*.

7. Works on "Europe" still tend to cover only the "West." See Wilson, *Food, Drink and Identity in Europe*, and Scholliers, *Food, Drink and Identity*. For a notable exception see Goldstein and Merkle, *Culinary Cultures of Europe*.

8. See Glants and Toomre's pioneering volume, *Food in Russian History and Culture*; Von Bremzen's amazing food memoir, *Mastering the Art of Soviet Cooking*; and Lakhtikova, Brintlinger, and Glushchenko's *Seasoned Socialism*.

9. See, for example, Bracewell, "Eating Up Yugoslavia." See also Caldwell, *Food and Everyday Life*.

10. See, for example, Pozharliev's comprehensive and philosophically rich work *Filosofiia na khraneneto*.

11. See the many works of Stefan Detchev, Raĭna Gavrilova, and Albena Shkodrova in Bulgarian and English, including Detchev, "'Shopska salata'"; Gavrilova, *Semeĭnata stsena*; and Shkodrova, *Sots gurme*.

12. See, for example, Işın, *Bountiful Empire*.

13. I will call this population "Bulgarians" for simplicity's sake, but it is important to note that a Bulgarian national consciousness was arguably consolidated only in the course of the twentieth century.

14. It was part of a complex of what Alexander Kiossev calls "self-colonization"—a self-imposed impulse to radically transform in the image of the West. Kiossev, "Self-Colonizing Metaphor."

15. This is a significant theme throughout my work. In particular see Neuburger, *Orient Within*. For the most notable work on the former Yugoslavia see Bakić-Hayden, "Nesting Orientalisms." See also Todorova, *Imagining the Balkans*, and Daskalov, *Mezhdu iztoka i zapada*.

16. On Russia see, for example, Neumann, *Russia and the Idea of Europe*; on Poland see Jedlicki, *Suburb of Europe*.

17. For the term "bioimaginary" I look to Steinberg, *Genes and the Bioimaginary*.

18. For an excellent overview of the history and meanings of the term "biopolitics" see Lemke, *Biopolitics*.

19. For this useful term and concept see Appadurai, *Social Life of Things*, 3–4.

20. On the complex phenomenon of nostalgia for communism in Eastern Europe with a focus on Bulgaria see Todorova and Gille, *Post-Communist Nostalgia*. For an excellent, nuanced discussion of the communist period in Bulgaria see Znepolski, *Bŭlgarskiiat komunizŭm*.

21. See Lih, *Bread and Authority*. See also Kitanina and Potolov, *Voina, khleb i revoliutsiia*.

22. For a recent scholarly examination of this phenomenon see, for example, Kulchytsky, *Famine of 1932–1933*.

23. "Lysenkoism" refers to the dominance in Soviet science (as dictated by Stalin) of the ideas of Trofim Lysenko. Namely, Lysenko assumed the heritability of "acquired" characteristics, as opposed to biological inheritance, which characterized Mendelian (after Gregor Mendel) genetics. After Stalin's death, Lysenko's approach to plant and animal science was shelved.

24. For a cross-Bloc comparison see Iordachi and Bauerkämper, *Collectivization of Agriculture*.

25. See, for example, Deutsch, *Food Revolution*, 75–78.

26. See, for example, Znepolski, *Bŭlgarskiiat komunizŭm*, 220.

27. For the official narrative on this improvement and statistics tracking it see Gocheva, *Razvitie na materialnoto blagosŭstoianie*. See also Iordanov, *Materialnoto-tekhnicheska basa na razvitoto*; and Lampe, *Bulgarian Economy*, 177, 192.

28. A wide variety of sources attest to the rise of a "middle class" of consumers in the region. On Bulgaria see Znepolski, *Bŭlgarskiiat komunizŭm*, 253, 274. On Romania see Chirot, "Social Change," 461. On Yugoslavia see Patterson, *Bought and Sold*.

29. Elias, *Civilizing Process*. For more on food and the civilizing process see Stephen Mennell, "On the Civilizing of Appetite."

30. See, for example, Scarboro, Mincyte, and Gille, *Socialist Good Life*.

31. See, for example, Kaser, *Household and Family*, 503.

32. See, for example, Riley, *Political History of American Food Aid*.

33. On tobacco production, consumption, and exchange see Neuburger, *Balkan Smoke*.

34. See, for example, Carolan, *Real Cost*; Castle and Goodman, *Meaty Truth*; Estabrook, *Tomatoland*.

35. "Most Obese Countries."

36. See Patel, *Stuffed and Starved*.

37. Von Bremzen, *Mastering the Art of Soviet Cooking*, 200–202.

38. See McGrath, *Food for Dissent*.

39. See, for example, Pells, *Not Like Us*, 296–306.

40. See, for example, Noncheva, *Winding Road*, 32.

41. As translated and cited by Ghodsee, "Red Nostalgia?," 33.

42. For a poignant look at some of the issues in post-socialist Bulgaria see Ghodsee, *Lost in Transition*.

1. By Bread Alone?

1. Belchov, "Iz borbite," 34. While Belchov puts the numbers in the hundreds, Georgi Vangelov argues for thousands of participants. See Vangelov, *Tiutiunorabotnitsi*, 99–100.

2. As a result of the incident, authorities increased bread rations by 125 grams. Vangelov, *Tiutiunorabotnitsi*, 100.

3. For more on this late war period see Crampton, *Bulgaria*, 211–14; on the Plovdiv incident see 306. As has been well documented, World War I afforded women expanded roles in societies, which had a significant effect in the war's immediate aftermath. See, for example, Sharp and Stibbe, *Aftermaths of War*.

4. On Russia, for example, see Choi Chatterjee's work on women's roles and later (mis)representations of those roles in the Russian Revolution and its aftermath. Chatterjee, *Celebrating Women*.

5. See also Volkov, "Aftermaths of Defeat."

6. See, for example, Laudan, *Cuisine and Empire*, 35–36.

7. Women's food riots have been deeply embedded in the "moral economy" of modern political revolutions. See Thompson, "Moral Economy," and Smith, "Food Rioters," 3–5, 26–29.

8. See, for example, Wood, *Baba and the Comrade*, and Lih, *Bread and Authority*.

9. Fernández-Armesto, *Civilizations*, 80.

10. Most Balkan peasants produced grains on small private plots, while Turkic-Ottoman landlords dominated the larger tracts of arable land in the lowland granaries. See, for example, İnalcık and Quataert, *Economic and Social History*, 180; and Palairet, *Balkan Economies*, 39.

11. Pavlov, *Prisŭstvia na khraneneto*, 8–9.

12. Pavlov, 107. MacDermott, *Bulgarian Folk Customs*, 61–62, 112–13.

13. Gavrilova, *Koleloto na zhivota*, 82.

14. For an observation of this in the interwar period see, for example, Leslie, *Where East Is West*, 28, 254.

15. Raĭkin, *Rebel with a Just Cause*, 101.

16. Pavlov, *Prisŭstvia na khraneneto*, 10–11. On the presence of urban bakers' guilds see Semov and Iankova, *Bŭlgarskite gradove*, 219, 339, 389. On bread preparation see also Kaneva-Johnson, *Melting Pot*, 231.

17. Gavrilova, *Koleloto na zhivota*, 83–85, 107.

18. For a fascinating exploration of the *burek* (or *biurek, banitsa*) in the post-Yugoslav context see Mlekuž, *Burek*.

19. See, for example, Karaosmanoğlu, "Cooking the Past."

20. Slaveĭkov, *Gotvarska kniga* (1870). For a more recent edition see Slaveĭkov, *Gotvarska kniga na Diado Slaveĭkov* (2015).

21. Slaveĭkov, *Gotvarska kniga* (2015), 57–58.

22. Slaveĭkov (2015), 45–47, 27–32.

23. Detchev, "Mezhdu visshata osmanska kukhnia."

24. Botev, *Sŭbrani sŭchineniia*, 6:57.

25. Crampton, "Turks in Bulgaria," 45.

26. Raĭkin, *Rebel with a Just Cause*, 29.

27. Todorov, "Deĭnostta," 36.

28. One decare equals ten acres. This emigration reduced the numbers of Turks from 26 percent of Bulgaria's population in 1878 to 14 percent in 1900. Crampton, "Turks in Bulgaria," 47.

29. Andreev, *Istoriia*, 65.

30. Crampton, *Bulgaria*, 209.

31. Tashev, *Khranene*.

32. The idea of a particular Russian (and East European) attraction to socialism as a consequence of and antidote to "backwardness" is by no means new, but it still bears repeating. For a recent exploration of this issue across the Balkans see Dimou, *Entangled Paths*.

33. "Bulgarian" is in quotation marks here because the identity of local Orthodox Slavs in that period, as now, was hotly disputed. Many have claimed that these populations are ethnically "Macedonian."

34. For a fascinating (leftist) view of these wars and this region "on the ground" see Trotsky, *Balkan Wars*, 399.

35. League of Nations, *European Conference*, 13.

36. On the implications of Bulgarian debt during this period see Lampe and Jackson, *Balkan Economic History*, 377.

37. For a discussion of food shortages in the region during these wars see Trotsky, *Balkan Wars*, 126–27, 255, 393.

38. See Trentmann and Just, *Food and Conflict*, 1.

39. Trentmann and Just, 1–2.

40. For a comprehensive study of the issue of bread and revolution in Russia see Lih, *Bread and Authority*.

41. Stevenson, *With Our Backs to the Wall*, 2.

42. Crampton, *Bulgaria*, 217.

43. Kastelov, *Bŭlgariia*, 53, 113.

44. To avoid confusion I am using "Bulgarian Communist Party" even though that name was not used until 1919. In this period the party was called the "Narrows," the Bolshevik-allied branch of the Bulgarian Workers Social Democratic Workers Party.

45. Stevenson, *With Our Backs to the Wall*, 407.

46. *Pod znamenem Oktiabria*, 2:49–50.

47. *Pod znamenem Oktiabria*, 2:51.

48. Kastelov, *Bŭlgariia*, 114–15.

49. *Pod znamenem Oktiabria*, 1:103.

50. *Pod znamenem Oktiabria*, 2:145.

51. The rule of Stamboliski is extremely controversial in the historical literature. For an overview see Bell, *Peasants in Power*. For a recent Bulgarian assessment see Daskalov, *Bŭlgarskoto obshtestvo*, 1:195.

52. Groueff, *Crown of Thorns*, 75.

53. Crampton, *Eastern Europe*, 120.

54. Seppain, *Contrasting US and German Attitudes*, 10.

55. Riley, *Political History*, 33–58.

56. Irwin, "Taming Total War," 123. See also Veit, *Modern Food*, 76.

57. See Patenaude, *Big Show*. See also Riley, *Political History*, 60–61.

58. Irwin, "Taming Total War," 23.

59. Petrov, *Agrarnite reformi*, 81–83.

60. Lampe and Jackson, *Balkan Economic History*, 185.

61. Tsentralen Dŭrzhaven Arkhiv [hereafter TsDA], fond [hereafter f.] 77, opis [hereafter op.] 1, arkhivna edinitsa [hereafter a.e.] 16, list [hereafter l.] 14.

62. See, for example, Zlatarov, *Osnovi na naukata*. The 2016 second edition added the subtitle *Lektsi, dŭrzhani prez uchebnata 1920–1 godina v Sofiiski Universitet*.

63. See Asen Zlatarov and Ivan Mitev, "Khrana v domashna biudzhet na Bŭlgarina," Plovdiv Okrŭzhen Dŭrzhaven Arkhiv [hereafter PODA], f. 1811, op. 1, a.e. 48, ll. 1–11. These studies built in part on interwar League of Nations–funded studies on food in Europe. See League of Nations, *European Conference*, 2.

64. PODA, f. 1811, op. 1, a.e. 48, l. 11.

65. Zlatarov, *Osnovi na naukata* (2016), 265.

66. Zlatarov (2016), 273–75.

67. Zlatarov, *Izbrani proizvedeniia*, 291.

68. Nabhan, *Where Our Food Comes From*.

69. Soyfer, *Lysenko*.

70. Pringle, *Murder of Nikolai Vavilov*, 2.

71. Drews, "Zashto ne deistvuvashe vegetarianstvoto," 133. See also Dande-Vansel, "Belite brashna," 91.

72. According to Pavel Biryukov, when Lev Tolstoy left Yasnaya Polyana (his estate) shortly before his death in 1910, his intention was to travel to Bulgaria to meet with Petŭr Dŭnov.

73. In Georgi Shopov's memoir of his travels to Yasnaya Polyana in Russia in 1913, he reports that numerous associates of Tolstoy told him that Tolstoy was on his way to Bulgaria in his dying days, presumably to visit the Bulgarian Tolstoyans. Shopov, *Na gosti*, 19, 29. Russian sources are less clear on this; according to one scholarly account, however, Tolstoy, though unsure of his destination, had written "somewhere abroad . . . perhaps to Bulgaria . . . or to Novocherkassk." See Bunin, *Liberation of Tolstoy*, 9.

74. See Kostentseva, *Moiat roden grad*, 183. For a list of some of his most influential followers see "Spiritual Master Peter Deunov."

75. Kostentseva, *Moiat roden grad*, 183.

76. See, for example, Douno, *Food and Water*, 40–41.

77. Douno, 40–41.

78. Douno, 41.
79. Douno, 46–47.
80. Douno, 48.
81. Georgiev, *Bial ili cher khliab*.
82. Georgiev, 3–4.
83. Georgiev, 5, 8.
84. Georgiev, 3–4.
85. Georgiev, 23, 39. On Nazi Germany and bread see Proctor, *Racial Hygiene*, 236.
86. Georgiev, *Bial ili cher khliab*, 36; Proctor, *Racial Hygiene*, 237.
87. See, for example, Gerhard, *Nazi Hunger Politics*, 35.
88. See, for example, Burlikov, "Blagosŭstoianie i khranene," 122–23.
89. Ganov, "Prekhranate na natsiata," 121.
90. See Crawford, *Economic Vulnerability*, 67; Castro, *Geography of Hunger*, 249.
91. See, for example, Shub and Warhaftig, *Starvation over Europe*, 8–26.
92. Collingham, *Taste of War*, 69.
93. Castro, *Geography of Hunger*, 248–49.
94. Shub and Warhaftig, *Starvation over Europe*, 8, 35–37, 88.
95. Lampe, *Bulgarian Economy*, 112–13.
96. Brandt, *Management of Agriculture and Food*, 211.
97. For a general discussion of Slavs under the Nazis see Connelly, "Nazis and Slavs." For an accounting of rations see Shub and Warhaftig, *Starvation over Europe*, 88.
98. Lampe, *Bulgarian Economy*, 81–82.
99. I borrow this turn of phrase from the classic work of Nissan Oren, *Revolution Administered*.
100. Semerdzhiev, *Narodniiat sŭd*, 330–31.
101. PODA, f. 1812k, op. 1, a.e. 16, l. 38.
102. TsDA, f. 77, op. 1, a.e. 22, l. 107.
103. TsDA, f. 77, op. 1, a.e. 16, l. 14; TsDA, f. 77, op. 1, a.e. 22, l. 170.
104. TsDA, f. 77, op. 1, a.e. 16, l. 14; TsDA, f. 77, op. 1, a.e. 22, l. 170.
105. Tashev, *Khranene*, 32.
106. Riley, *Political History*, 103.
107. Riley, 121.
108. Riley, 144.
109. United Nations Interim Commission, *Work of FAO*.
110. United Nations Interim Commission, 146.
111. See Tashev, *Khranene*, 40.
112. The most nuanced source on collectivization in Bulgaria is Gruev, *Preorani slogove*. For a more negative view of collectivization see Ĭosifov, *Totalitarizmŭt*.
113. Crampton, *Bulgaria*, 329. One hectare is about 2.5 acres.
114. Ĭosifov, *Totalitarizmŭt*, 94.
115. Ĭosifov, 97–98.
116. Lampe, *Bulgarian Economy*, 146.
117. Ĭosifov, *Totalitarizmŭt*, 103.
118. Though Turks were willing participants in the mass exodus, outside sources generally label it an "expulsion" because of the haste and intent behind the operation. The destabilization of pro-NATO Turkey was one probable objective, along with collectivization.

119. Tepev, *Turskoto naselenie*, 69.

120. Stillman, "Collectivization," 84.

121. See, for example, discussion on this in the section on Bulgaria in Armstrong, *Tito and Goliath*, 218.

122. Judt, *Postwar*, 163.

123. See, for example, McCauley, *Khrushchev*, xi.

124. See, for example, Embree, *Soviet Union*, 165. See also Ploss, *Conflict and Decision Making*.

125. See Paulina Bren and Mary Neuburger, introduction to *Communism Unwrapped*, 8–9.

126. McCauley, *Khrushchev*, xii.

127. Riley, *Political History*, 214–18.

128. Khrushchev and Khrushchev, *Memoirs*, 3:143, 148.

129. Tomov, *Neshta*, 3–4.

130. For more on rational consumption see Crowley and Reid, "Style and Socialism," 10.

131. Tomov, *Neshta*, 7–9, 70 (on the problems of white bread), 92 (on alcohol).

132. PODA, f. 1, op. 1, a.e. 4, l. 19.

133. Dimitrov, Raĭchev, and Stefanov, *Mekhanizatsiia i avtomatizatsiia*, 3.

134. Tashev, *Khranene*, 32.

135. TsDA, f. 172, op. 3, a.e. 153, ll. 2–4.

136. TsDA, f. 172, op. 3, a.e. 153, l. 9.

137. For the best history of American white bread see Bobrow-Strain, *White Bread*.

138. Bobrow-Strain, ix.

139. Bobrow-Strain, x.

140. Golovinski, *Osŭshtestvenata mechta*, 94.

141. See, for example, Tashev, *Khranene*, 25.

142. Tashev, 44.

143. Tashev, 6, 11. See also Golovinski, *Osŭshtestvenata mechta*, 94.

144. Petrov et al., *Bŭlgarska natsionalna kukhnia*, 7.

145. Tashev, *Khranene*, 31, 44–45.

146. Tashev, 45.

147. Tashev, 25–41.

148. "Economic Growth Brisk in Red Bloc," *New York Times*, March 20, 1967, 41. Statistics are drawn from the UN Economic Commission for Europe.

149. Radio Free Europe, *Radio Free Europe Research*, 305.

150. Johnson, *Gay Bulgaria*, 34.

151. See, for example, Frankel, "Red Countries Seek U.S. Wheat."

152. See, for example, *Hearings before the Committee on Foreign Relations*, 110.

153. Kennedy, *Kennedy Presidential Press Conferences*, 563.

154. *Hearings before the Committee on Foreign Relations*, 35.

155. Kennedy, "Letter to the President of the Senate."

156. See, for example, Wädekin, Eugen, and Jacobs, *Agrarian Policies*, 126–27.

157. See, for example, Trager, *Great Grain Robbery*, 233.

158. A 1974 world food conference convened in Rome under the auspices of FAO. This brought public attention, debate, and more studies on the issue of global

hunger and global grain supplies, with exchanges of information on and plans for the global food system.

159. See, for example, Desai, *Estimates*, 7–8.

160. Andelman, "An Oil Shock of Their Own."

161. TsDA, f. 130, op. 23, e.a. 300, l. 6.

162. Tashev, *Khranene*, 45.

163. Chortanova and Dzhelepov, *Nashata i svetovnata kukhnia*, 83.

164. See, for example, Kauffman, *Hippy Food*.

165. Dilchev, *Zagadŭchnata smŭrt*, 59.

166. Dimkov, *Bŭlgarska narodna meditsina*, 5, 9, 15.

167. Dimkov, 90, 94.

168. Petrov et al., *Bŭlgarska natsionalna kukhnia*, 8.

169. Petrov et al., 7.

170. Dzhelepov and Belorechki, "Dietichno khranene," 326.

171. See, for example, Zhekova, *Shto e ratsionalna khranene*, 4–5.

172. Todorov, *Golemiiat khliab*, 5–6.

173. Cochrane, *Hard Currency Constraints*, 5.

174. Centre for Co-operation with Non-Members, *Review of Agricultural Policies*, 51–52.

175. See, for example, Luif, *Security*, 215.

176. Centre for Co-operation with Non-Members, *Review of Agricultural Policies*, 101.

177. Collar, "Bread," 500.

2. Vegetarian Visions and Meatopias

1. On the phenomenon of communist elite hunting in Hungary see Peteri, "Nomenklatura."

2. Murdzhev, *Taka gi vidiakh*, 30–32.

3. For an amazingly detailed story about Ceaușescu and his hunting practices see Quammen, "Bear Slayer."

4. Stoĭchkov, *Lovnite retsepti*, 3. "Dancho" is the diminutive for "Iordan," and "Bai" is a common pre-communist form of address with a folksy tone akin to "uncle."

5. See Ogle, *In Meat We Trust*.

6. See, for example, Zaraska, *Meathooked*, 115–17.

7. See, for example, LeBlanc, "Ethics and Politics."

8. See my detailed discussion of this phenomenon and contested process, particularly in relation to Muslim minorities, in Neuburger, *Orient Within*. On "Orientalization" of the Balkans see also Bakić-Hayden, "Nesting Orientalisms." See also Maria Todorova's important work on the related phenomenon of "Balkanism" in *Imagining the Balkans*.

9. See, for example, Toussaint-Samat, *History of Food*, 75–77.

10. Stanford, *Hunting Apes*, 75, 210.

11. Wrangham, *Catching Fire*, 53–54.

12. See, for example, Stuart, *Bloodless Revolution*; Spencer, *Heretic's Feast*; Walters and Portmess, *Ethical Vegetarianism*.

13. See, for example, Fernández-Armesto, *Civilizations*, 455.

14. Zaraska, *Meathooked*, 200.

15. See Jezernik, *Wild Europe*, 47–55, 147–70. Foreigners were wont to label the region "the Orient" well after 1878, and indeed into the interwar period. See, for example, Fermor, *Broken Road*, 3.

16. St. Clair and Brophy, *Residence in Bulgaria*, 45. They also noted that meat and bones were often used in healing and soothsaying; see 36–37, 43–44.

17. St. Clair and Brophy, 110.

18. Barkley, *Bulgaria before the War*. See, for example, Fox, *Bulgaria*, 121.

19. Barkley, *Bulgaria before the War*, 36.

20. Barkley, 37.

21. See, for example, Pavlov, *Prisŭstviia na khraneneto*, 124–28.

22. Gavrilova, *Koleloto na zhivota*, 79–81.

23. Pavlov, *Prisŭstviia na khraneneto*, 137.

24. For an excellent discussion of this in the Russian context, through the lens of literary texts, see LeBlanc, *Slavic Sins*.

25. MacDermott, *Bulgarian Folk Customs*, 86.

26. Gavrilova, *Koleloto na zhivota*, 71–72.

27. Pavlov, *Prisŭstviia na khraneneto*, 78–79.

28. Slaveĭkov, *Gotvarska kniga* (2015), 102.

29. Animal slaughter was generally seasonal, peaking in the fall after animals had been fattened by summer pasturing and females were done lactating. Fall slaughter also meant that animals did not have to be fed over the winter or moved to summer grazing areas. Tsachev, Ĭoncheva, and Mladenova, *Ot vŭrkha na XX vek*, 41.

30. Tsachev, Ĭoncheva, and Mladenova, 39.

31. Tsachev, Ĭoncheva, and Mladenova, 80.

32. Tsachev, Ĭoncheva, and Mladenova, 174.

33. Tsachev and Stoĭchev, *Mesopromishlenostta*, 14.

34. Sharkey, *History of Muslims, Christians, and Jews*, 87; Tsachev, Ĭoncheva, and Mladenova, *Ot vŭrkha na XX vek*, 42.

35. See, for example, Tsachev, Ĭoncheva, and Mladenova, *Ot vŭrkha na XX vek*, 14.

36. For the meat preferences of a variety of revival and post-revival figures see Velichkov, *Kakvo khapnakha*, 98–123.

37. Pavlov, *Prisŭstviia na khraneneto*, 48.

38. Slaveĭkov, *Gotvarska kniga* (1870), 9–15, 23–24.

39. Slaveĭkov (1870), 83–91.

40. Slaveĭkov (1870), 102.

41. Slaveĭkov (1870), 102.

42. See, for example, Botev, *Sŭbrani sŭchineniia*, 2:71; Rakovski, *Izbrani sŭchineniia*, 47; and Karavelov, "Zapiski za Bŭlgariia i Bŭlgarite," 171–73, 221.

43. Tsachev, Ĭoncheva, and Mladenova, *Ot vŭrkha na XX vek*, 51.

44. Tsachev, Ĭoncheva, and Mladenova, 78.

45. Crampton, *Bulgaria*, 158–59.

46. Kostentseva, *Moiat roden grad*, 63–65; Leslie, *Where East Is West*, 63.

47. See Velichkov, *Kakvo khapnakha*, 7–9.

48. Velichkov, *Kakvo khapnakha*, 5–6; Leslie, *Where East Is West*, 155.

49. Here I use "nostalgia" in what Svetlana Boym calls the "reflective" as opposed to "restorative" sense. That is, I am not arguing that Vazov or others in the post-Ottoman (post-1878) decades were in any way attempting to restore or return to the way of life under the Ottomans by evoking food or other sensory images. Rather, Vazov was one of many for whom food elicited a certain "longing" for the fragments (and flavors) of the Ottoman years. See Boym, *Future of Nostalgia*, 50.

50. Vazov, *Under the Yoke*, 76.

51. Vazov, *Great Rila Wilderness*, 96.

52. Vazov, 36.

53. Vazov, 36–43.

54. Vazov, 87, 103, 121.

55. Tolstoy, *First Step*, 38. On the virtues of the peasant fast see 43.

56. Tolstoy, 52–57.

57. Ogle, *In Meat We Trust*, 45.

58. Shprintzen, *Vegetarian Crusade*, 156.

59. On the significance of Eastern travelers to the West see Bracewell and Drace-Francis, *Under Eastern Eyes*.

60. See also the sole translation to date: Konstantinov, *To Chicago and Back*, 62.

61. Konstantinov, 37.

62. Konstantinov, *Do Chikago i nazad*, 79.

63. See, for example, Brockett, *Bogomils*.

64. Kostentseva, *Moiat roden grad*, 183. Stoian Vatralski, "Koi i kakvi sa Belite bratia," 192, 204.

65. Vatralski, "Koi i kakvi sa Belite bratia," 204.

66. See Kostentseva, *Moiat roden grad*, 183.

67. Konstantinov, *L. N. Tolstoĭ*.

68. Edgerton, "Social Influence," 128–29.

69. Kiumzhiev, "Shtadete zhivota!," 4.

70. Lozinski, "Mesoiadie i prestŭpnost," 154.

71. Edgerton, "Social Influence," 129.

72. For other publications see, for example, Andreĭchin, *Ezikŭt na vegetarianstvo*; and Dosev, *Etika na khranata*.

73. Edgerton, "Social Influence," 130–32.

74. Dosev, *Etika na khranata*, 51–59.

75. Dosev, 91–93.

76. Ivan Kiumzhiev, "Kŭm delo!," 1.

77. Lozinski, "Mesoiadie i prestŭpnost," 154.

78. Lozinski, 154.

79. Lozinski, 154.

80. "Peti kongres na Vegetarianski siŭz," 5.

81. See, for example, Albrecht, "Vegetarianstvo," 41.

82. Andreĭchin, "Kŭm dŭlboko osŭznavane," 122.

83. Andreĭchin, 122.

84. St. Gavriĭski, "Otraviane," 212–15.

85. See, for example, Stanchev, "Nauka na khranata," 180–81. See also V. M., "Protein," 182–83.

86. See Stanchev, "Khrana i khranitelnost," 35.

87. Ivanov, "Tagore in Bulgaria," 329–31.

88. See, for example, Zlatarov, *Osnovi na naukata* (1921).

89. Golovinski, *Osŭshtestvenata mechta*, 106.

90. Tsachev, Ĭoncheva, and Mladenova, *Ot vŭrkha na XX vek*, 105.

91. On consumption see Tsachev, Ĭoncheva, and Mladenova, *Ot vŭrkha na XX vek*, 110. For export numbers see 125.

92. Shkodrova, *Sots gurme*, 40.

93. United States Mission, Sofia, Bulgaria, December 26, 1946, "The Food and Agricultural Prospects in Bulgaria 1945–48," United Nations Archive, UNRRA Collection, 10–11. For more on World War II see chapter 1.

94. "Meat Rations and Butcher Stores at Stalin," May 2, 1952 [electronic record], HU OSA 300-1-2-19112, Records of Radio Free Europe / Radio Liberty Research Institute: General Records: Information Items, Open Society Archives at Central European University, Budapest, http://hdl.handle.net/10891/osa:3a0c3152-55b3-41e2-b8a2-1c006e6c375d.

95. Tsachev, Ĭoncheva, and Mladenova, *Ot vŭrkha na XX vek*, 94–96.

96. TsDA, f. 172, op. 3, a.e. 149, ll. 1–41; TsDA, f. 172, op. 3, a.e. 150, ll. 1–11.

97. See, for example, "Pork and Bacon."

98. Tsachev and Stoĭchev, *Mesopromishlenostta*, 63, 118.

99. TsDA, f. 172, op. 3, a.e. 150, l. 3.

100. Tsachev, Ĭoncheva, and Mladenova, *Ot vŭrkha na XX vek*, 120.

101. TsDA, f. 1B, op. 509, a.e. 5, l. 51, 1962.

102. For more on the assimilation of Turks in the 1980s see Neuburger, *Orient Within*.

103. Golovinski, *Osŭshtestvenata mechta*, 94.

104. See, for example, Tashev, *Khranene*, 25.

105. See, for example, Tashev, *Khranene*, 6, 11. See also Golovinski, *Osŭshtestvenata mechta*, 94.

106. Petrov et al., *Bŭlgarska natsionalna kukhnia*, 7.

107. Demireva, *Izmeneniia v khraneneto*, 63.

108. Tashev, *Khranene*, 85.

109. See Tsachev, Ĭoncheva, and Mladenova, *Ot vŭrkha na XX vek*, 109–10.

110. Golovinski, *Osŭshtestvenata mechta*, 102.

111. Tashev, *Khranene*, 11.

112. Naĭdenov and Chortanova, *Rŭkovodstvo*, 6.

113. Naĭdenov and Chortanova, 11.

114. Note: I am translating the author's Bulgarian quotes of Engels into English rather than returning to the original. Naĭdenov and Chortanova, *Rŭkovodstvo*, 32–33.

115. Naĭdenov and Chortanova, *Rŭkovodstvo*, 32.

116. If this is not self-evident, Petŭr Saraliev's *Gotvarska kniga za mŭzhe* (A cookbook for men), published in 1984, makes it abundantly clear.

117. For an excellent article on the evolution of the role of the cookbook see Shkodrova, "From Duty to Pleasure." For a more complete discussion of cookbooks in socialist Bulgaria see Shkodrova, *Rebellious Cooks*.

118. Although there is also no section for vegetarian or *postni* foods, the word *postni* is randomly used as a heading on a list of vegetable dishes that gives nutritional values. Naĭdenov and Chortanova, *Nasha kukhnia*, 88.

119. Naĭdenov and Chortanova, 39.

120. Sotirov, *Sŭvremenna kukhnia*.

121. Shkodrova, *Sots gurme*, 149.

122. Sŭbev, *90 godini organizirano turistichesko dvizhenie*, 11. While this was a far lower number than socialist Yugoslavia's reported 2.6 million tourists in 1965 and 8.4 million in 1985, it was still impressive given Bulgaria's size and resources. See Turncock, *East European Economy*, 45.

123. Tsachev and Stoĭchev, *Mesopromishlenostta*, 55.

124. Haskell, *Heroes and Roses*, 47.

125. For an interesting look at *kebabche* see Georgieff, "Rise and Fall."

126. Petrov et al., *Bŭlgarska natsionalna kukhnia*, 348.

127. Petrov et al., 6–7. This 100 percent benchmark was met not solely through meat but also, as explored in the next chapter, through major increases in dairy (along with eggs).

128. Petrov et al., *Bŭlgarska natsionalna kukhnia*, 11–12.

129. Petrov et al., 325.

130. See, for example, Iacobbo and Iacobbo, *Vegetarian America*, 169.

131. Tchakarov, *Second Floor*, 142.

132. For the best study of this see Dragostinova, *Cold War*, 162–221.

133. Aleksandrov, *Kultura i lichna vlast*, 69.

134. Blagov, *Zagadkata Liudmila Zhivkova*, 74.

135. Dragostinova, *Cold War*, 173–77. See also Petrov, "Rose and the Lotus."

136. See, for example, Zhivkova, *Lyudmila Zhivkova*, 118, 309.

137. Blagov, *Zagadkata Liudmila Zhivkova*, 127.

138. Tchakarov, *Second Floor*, 138.

139. Tchakarov, 75.

140. Blagov, *Zagadkata Liudmila Zhivkova*, 148.

141. Some sources doubt her commitment to Marxism, if not her sanity, in the final years of her life. See, for example, Tchakarov, *Second Floor*, 140–41.

142. Blagov, *Zagadkata Liudmila Zhivkova*, 103.

143. Dilchev, *Zagadŭchnata smŭrt*, 59.

144. Dimkov, *Bŭlgarska narodna meditsina*, 16.

145. Belorechki, *Vegetarianstvo i surovoiadstvo*.

146. Dilchev, *Zagadŭchnata smŭrt*, 8.

147. For details on rumors that Zhivkova's eating regimen weakened or even killed her see Aleksandrov, *Kultura i lichna vlast*, 69, 107. Also see Tchakarov, who claims that Zhivkova deteriorated in her final years, undermining her "frail constitution as she starved herself": *Second Floor*, 140.

148. Zlatarov, *Osnovi na naukata za khraneneto* (2016), 271–72. This is a later edition of the 1981 version.

149. Tsachev, Ĭoncheva, and Mladenova, *Ot vŭrkha na XX vek*, 150–52.

150. FAOSTAT, Food and Agriculture Organization of the United Nations, http://www.fao.org/faostat/en/.

3. Sour Milk

1. Dzhelepov and Belorechki, "Dietichno khranene," 326.

2. Mendelson, *Milk*, ix.

3. Metchnikoff, *Prolongation of Life*.

4. Dzhelepov and Belorechki, "Dietichno khranene," 302.

5. Dzhelepov and Belorechki, 327.

6. "Metchnikoff Confirmed in His Theory of Long Life," *New York Times*, January 21, 1912, 13.

7. The article cites, for example, Douglas, *Bacillus*.

8. See, for example, Debré, *Louis Pasteur*, 82–115.

9. See, for example, Valenze, *Milk*, 226–27.

10. Metchnikoff, *Prolongation of Life*.

11. Metchnikoff, 182.

12. Nauchno-konsultativen sŭvet, *Bŭlgarskoto ime*, 98.

13. Fondatsiia "Dr Stamen Grigorov," *V nachaloto*, 97.

14. Slosson, "Twelve Major Prophets," 1235.

15. See, for example, "Metchnikoff's Theory," and Kendall, "Recent Developments."

16. "Science and Sour Milk," (London) *Times*, April 29, 1910.

17. See, for example, Valenze, *Milk*, 188. On a similar process in the UK see Nimmo, *Milk, Modernity*.

18. Wiley, *Re-imagining Milk*.

19. Herald of Health, "Sour Milk Fetish," 262.

20. Shepard, "What the Doctors Say," 118.

21. "Uses of Sour Milk."

22. Kellogg, *Battle Creek Sanitarium*, 121, 131.

23. Schwarcz, *Apple a Day*, 212.

24. Mendelson, *Milk*, 150.

25. Mendelson, 7.

26. See, for example, Kondratenko, *Bŭlgarsko kiselo mliako*, 7.

27. On the longer history of Thracians in the Bulgarian historical imagination see Marinov, "Ancient Thrace," 75–112. For a reference to Thrace in the literature on yogurt see Katrandzhiev, *Bŭlgarskoto kiselo mliako*, 8–9.

28. Katrandzhiev, *Bŭlgarskoto kiselo mliako*, 10.

29. Ivanov, *Kiselo mliako*, 18. See also Gavrilova, *Koleloto na zhivota*, 88.

30. Karavelov, *Zapiski*, 156–59.

31. See Gavrilova, *Koleloto na zhivota*, 79–80.

32. Khadzhiĭski and Khadzhiĭska, *Bit i dushevnost*, 223.

33. Kostentseva, *Moiat roden grad*, 66.

34. See, for example, Karaosmanoğlu, "Cooking the Past."

35. For the most detailed recounting of Ottoman food see Işın, *Bountiful Empire*. On nineteenth-century Bulgarian urban food see Gavrilova, *Koleloto na zhivota*, 81.

36. See, for example, Işın, *Bountiful Empire*, 25, 133, 142.

37. Slaveĭkov, *Gotvarska kniga* (1870). For a more recent edition see also Slaveĭkov, *Gotvarska kniga na Diado Slaveĭkov* (2015).

38. Slaveĭkov, *Gotvarska kniga* (2015), 64.

39. For the first iteration see Crampton, *Bulgaria*, 159. For the second see Elenkov, "Versii za bŭlgarskata identichnost," 14.

40. Although Michael Palairet, John Lampe, and others argue that the "penetration" of the West was relatively insignificant during this period, it is clear that the *perceived* impact was still significant. See Palairet, *Balkan Economies*, 175–78; and Lampe, "Imperial Borderlands," 200–202.

41. See Stoilova, "Producing Bulgarian Yoghurt," 57–80.

42. See Asen Zlatarov and Ivan Mitev, "Khrana v domashna biudzhet na Bŭlgarina," PODA, f. 1811, op. 1, a.e. 48, ll. 1–11. These studies built in part on interwar League of Nations–funded studies on food in Europe. See League of Nations, *European Conference*, 2.

43. For explicit (and repeated) reference to this see Kantardzhiev, *Mlekarski narechnik*, 3, 109, 202.

44. *Nova gotvarska kniga*, 3.

45. *Nova gotvarska kniga*, 77.

46. See Stoilova, "From a Homemade to an Industrial Product."

47. Kantardzhiev, *Mlekarski narechnik*, 154–200.

48. Stoilova, "Producing Bulgarian Yoghurt," 93.

49. Stoilova, 43–47, 92.

50. According to Kantardzhiev, there were two hundred in Sofia as of 1930. *Mlekarski narechnik*, 197.

51. Kantardzhiev, *Mlekarski narechnik*, 195.

52. Kantardzhiev, 197.

53. Popdimitrov, *Bŭlgarsko kiselo mliako*, 9–10.

54. Popdimitrov, 20.

55. Popdimitrov, 22–25.

56. Popdimitrov, 5.

57. Popdimitrov, 30, 87.

58. Partŭchev, *Nashite stoletnitsi*.

59. Partŭchev, 7.

60. Partŭchev, 18.

61. See, for example, Sheimanov, "Preobrazhenie na Bŭlgariia," 267.

62. Dimov, *Mlekarstvo*, 5.

63. See, for example, Dimov, *Mlekarstvo*; Minchev, *Kratko praktichesko rŭkovodstvo*; Emanuilov, *Veterinarno-sanitarna ekspertiza*.

64. Dimov, *Mlekarstvo*, 4.

65. Naĭdenov and Zakhariev, *Prouchvane*, 4.

66. Stoilova, "Producing Bulgarian Yoghurt," 109–10.

67. Stoilova, 121.

68. Naĭdenov and Zakhariev, *Prouchvane*, 6.

69. Popdimitrov, *Bŭlgarsko kiselo mliako*, 31–35.

70. Stoilova, "Producing Bulgarian Yoghurt," 111.

71. Stoilova, 115–19.

72. Nauchno-konsultativen sŭvet, *Bŭlgarskoto ime*, 98.

73. Belorechki and Dzhelepov, *Mlechna kukhnia*, 18–19.

74. Kondratenko, *Bŭlgarsko kiselo mliako*, 249.

75. For an overview of the controversies see Podolsky, "Cultural Divergence." For the study that claimed that *Bacillus bulgaricus* did not survive in the gut (while acidophilus did) see Kulp and Rettger, "Comparative Study."

76. Mendelson, *Milk*, 150.

77. Georgiev, *Dŭlgoletieto*, 32.

78. Georgiev, 20.

79. Katrandzhiev, *Bŭlgarskoto kiselo mliako*, 28.

80. Katrandzhiev, 32.

81. Katrandzhiev, 5.

82. Naĭdenov and Chortanova, *Rŭkovodstvo*; Naĭdenov and Chortanova, *Nasha kukhnia*.

83. Naĭdenov and Chortanova, *Nasha kukhnia*, 40.

84. Naĭdenov and Chortanova, 115.

85. Naĭdenov and Chortanova, 153, 184, 213.

86. See Neuburger, *Orient Within*.

87. Petrov et al., *Bŭlgarska natsionalna kukhnia*, 7.

88. Petrov et al., 23.

89. Cheeses, in particular soft (feta-style) types, but also aged (Kashkaval) varieties, were also important ingredients of this national cuisine. Like yogurt, cheese was an important source of protein, but it lacked the imputed health value grounded in probiotics and in some respects the culinary flexibility of yogurt.

90. Petrov et al., *Bŭlgarska natsionalna kukhnia*, 36.

91. Cholcheva, Chortanova, and Ilieva, "Umeem da gotvim," 415–16.

92. Cholcheva, Chortanova, and Ilieva, 41.

93. Cholcheva, Chortanova, and Ilieva, 8.

94. Markov, *Bulgarian Temptations*.

95. Markov, 22, 28.

96. Markov, 27.

97. Markov, 66.

98. Markov, 76.

99. See Nauchno-konsultativen sŭvet, *Bŭlgarskoto ime*, 121.

100. Nauchno-konsultativen sŭvet, 124. Meiji Milk still sells "plain" Bulgarian yogurt, which currently has a 48 percent market share and has government "approval" as "promoting digestive health."

101. Yotova, "'It Is the Bacillus,'" 175.

102. FAOSTAT, Food and Agriculture Organization of the United Nations, http://www.fao.org/faostat/en/#compare.

103. Megyeri, "Bulgarian Yoghurt Market Dynamics."

104. Yotova, "Reflecting Authenticity."

105. Konstantinova, *Zov za budnost*, 207.

106. See examples in these online "memories from the People's Republic of Bulgaria": "Kogato kiseloto mliako beshe kiselo, a mliakoto ot krava—I ch.," Socbg.com, December 19, 2013, http://socbg.com/2013/12/когато-киселото-мляко-беше-кисело-а-мл-2.html; Asen Ivanov, "Kogato kiseloto mliako i sireneto biakha istinski," Socbg.com, March 2019, http://socbg.com/2019/03/когато-киселото-мляко-и-сиренето-бяха.html.

107. "How to Piss Off a Bulgarian," *99 Lives* (blog), April 26, 2017, https://99livesblog.com/2017/04/26/how-to-piss-off-a-bulgarian/.

108. M. Shahbandeh, "U.S. Yogurt Market: Statistics & Facts," Statista, March 15, 2018, https://www.statista.com/topics/1739/yogurt/.

109. "Lawsuit: Yogurt Brand Has Nothing Greek about It," CNBC, June 20, 2014, https://www.cnbc.com/2014/06/20/lawsuit-yogurt-brand-has-nothing-greek-about-it.html.

4. "Ripe" Communism

1. On this and the place of fruits and vegetables in Bulgarian cuisine see Gavrilova, "Golden Fruits," 100.

2. See, for example, Laudan, *Cuisine and Empire*, 315–16.

3. For a recent work on how the tomato conquered the world see Hyman, *Tomato*. For country-specific works see Smith, *Tomato in America*; Gentilcore, *Pomodoro!* On peppers see, for example, Walton, *Devil's Dinner*, and Anderson, *Chillies*.

4. "Pork and Bacon."

5. Laudan, *Cuisine and Empire*, 161.

6. Artan, "Aspects."

7. See Gavrilova, "Golden Fruits."

8. According to a survey conducted in the period, about 70 percent of Bulgarians pickled vegetables, with only 12 percent practicing canning. Mocheva, *Selskoto zemedělsko domakinstvo*, 25.

9. See Detchev, "Shopska Salad," 276–77.

10. Gavrilova, *Koleloto na zhivota*, 94–95.

11. See Detchev, "'Liuti chushki,'" 56.

12. Detchev, 69–70.

13. Gavrilova, *Semeǐnata stsena*, 321.

14. Walton, *Devil's Dinner*, 135–42.

15. Detchev, "'Liuti chushki,'" 62.

16. See Konstantinov, *Bai Ganyo*, 53, 54.

17. Anderson, *Chillies*, 52. See also Lang, *Cuisine of Hungary*, 128.

18. Detchev, "Shopska Salad," 276.

19. Strauss, "Twenty Years in the Ottoman Capital," 252.

20. As Detchev argues, the taste for hot peppers in Bulgaria seems to have been initially concentrated in regions adjacent to Serbia and Hungary, indicating that even if Bulgarians brought peppers north, culinary tastes traveled both ways. Detchev, "'Liuti chushki,'" 70.

21. See, for example, Geshov, *Nashite gradinari druzhestva*, 1.

22. Gavrilova, *Semeǐnata stsena*, 321.

23. For an insightful analysis of this cookbook see Detchev, "From Istanbul to Sarajevo."

24. Slaveǐkov, *Gotvarska kniga na Diado Slaveǐkov*, 37–40, 79.

25. Gavrilova, *Semeǐnata stsena*, 257, 325.

26. See, for example, Karakasheva, *Gotvarska kniga*.

27. Gavrilova, *Semeǐnata stsena*, 333–37.

28. Kostentseva, *Moiat roden grad*, 76.

29. Kiril Popov, *Stopanska Bŭlgariia*, 145.

30. Popov, 205.

31. Castro, *Geography of Hunger*, 237.

32. League of Nations, *Problem of Nutrition*.

33. Popkin, "Nutritional Patterns."

34. See, for example, Zlatarov, "Gladuvashta Bŭlgariia," as cited in *Izbrani sŭchineniia*, 2:281–48.

35. Zlatarov, *Izbrani proizvedeniia*, 291.

36. Zlatarov, 305.

37. Zlatarov, *Osnovi na naukata*, 2nd ed., 273–75.

38. Zlatarov, *Izbrani sŭchineniia*, 2:321–22.

39. Zlatarov, *Osnovi na naukata*, 172.

40. See, for example, Bozhilova, "Novo poznanie," 60.

41. "Vegetarianska kukhnia," *Vegetarianski pregled* 12, no. 7 (March 1931): 109–10.

42. Detchev, "Shopska salata," 420–21.

43. "Zelenchutsi i salati," *Vegetarianski pregled* 8, no. 8–9 (April–May 1927): 194.

44. DAP, f. 1, op. 1, a.e. 4, l. 23.

45. DAP, f. 1, op. 1, a.e. 4, l. 23.

46. Gavrilova, *Semeĭnata stsena*, 323, 253.

47. See Castro, *Geography of Hunger*, 249.

48. Tomov, *Neshto koito traibva*, 3–4.

49. For more on rational consumption see Crowley and Reid, "Style and Socialism," 10.

50. Tomov, *Neshto koito traibva*, 7–9.

51. Tomov, 90.

52. Tomov, 90–91.

53. Tomov, 90.

54. PODA, f. 1, op. 1, a.e. 4, l. 19.

55. See chapter 2.

56. PODA, f. 1, op. 1, a.e. 4, l. 37.

57. United Nations Interim Commission, *Work of FAO*, 11–12.

58. TsDA, f. 77, op. 1, a.e. 22, ll. 2–21.

59. TsDA, f. 77, op. 1, a.e. 22, ll. 54–57.

60. TsDA, f. 172, op. 1, a.e. 61, l. 1.

61. TsDA, f. 172, op. 1, a.e. 61, l. 1.

62. TsDA, f. 172, op. 1, a.e. 61, ll. 6–7.

63. TsDA, f. 175, op. 1a, a.e. 50, ll. 41–42.

64. TsDA, f. 175, op. 1a, a.e. 50, ll. 41–42.

65. See, for example, Deutsch, *Food Revolution*, 75–78.

66. TsDA, f. 130, op. 23, a.e. 298, l. 10.

67. TsDA, f. 130, op. 23, a.e. 298, l. 11.

68. TsDA, f. 130, op. 23, a.e. 298, l. 12b.

69. See, for example, Ram and Yadava, *Genetic Resources*, 247. McLeod, Guttman, and Eshbaugh, "Peppers," 189.

70. TsDA, f. 130, op. 23, a.e. 298, ll. 11–13.

71. *Mezhdunarodna konferentsiia po vitaminite*, 21–23.

72. *Mezhdunarodna konferentsiia po vitaminite*, 25.

73. *Bulgaria Today* 9 (1971): 38.

74. *Bulgaria Today* 9 (1971): 39.

75. *Bulgaria Today* 9 (1971): 40–41.

76. Salisbury, "Bulgaria Seeks Trade."

77. Lyndon Baines Johnson Presidential Library, Personal Papers, Orville Freeman, box 35, Eastern Europe Trip, memo to Dean Rusk with details of trip, 5, and Byron Shaw report, 55, 62.

78. Currie, *Flying Scotsman*, 35.

79. See Cochrane, Schmitz, and Bojnec, "Agriculture," 48.

80. As cited in McCauley, *Khrushchev Era*, 106.

81. Khrushchev and Khrushchev, *Memoirs*, 3:448.

82. Khrushchev and Khrushchev, 3:448, 2:411.

83. Khrushchev and Khrushchev, 3:448.

84. Deutsch, *Food Revolution*, xvi–xvii.

85. Deutsch, 70.

86. Zaharieva, *Nine Rabbits*, 31.

87. Zaharieva, 31.

88. Zaharieva, 29–30.

89. Smollett, "Economy of Jars," 125–28.

90. Gavrilova, *Semeĭnata stsena*, 255; Jung, "From Canned Food," 32.

91. Khadzhiĭski et al., *Domashno konservirane*, 4–5. See also Bozukova and Zakhariev, *Domashno konservirane*.

92. Khadzhiĭski et al., *Domashno konservirane*, 14–15.

93. Jung, "From Canned Food," 30–32.

94. Organisation for Economic Co-operation and Development, *OECD Review*, 72.

95. Marinov, "Material."

96. Naĭdenov and Chortanova, *Nasha kukhnia*, 216.

97. Petrov et al., *Bŭlgarska natsionalna kukhnia*, 157–59.

98. Saraliev, *Gotvarska kniga za mŭzhe*.

99. Detchev, "Shopska salata," 278–80.

100. Kassabova, *Street without a Name*, 88.

101. Kassabova, 62.

102. See, for example, a discussion of this in Shkodrova, *Sots gurme*, 9.

103. See Jung, "From Canned Food," 33–36.

104. See, for example, Estabrook, *Tomatoland*.

5. Wine and Dine

1. Zaĭkov et al., *Kniga za vinoto*, 16.

2. Zaĭkov et al., 20.

3. For the best description of the jubilee see Dragostinova, *Cold War*, 40–48.

4. On Bulgaria see Znepolski et al., *Bulgaria under Communism*. For a Bloc-wide analysis of nationalism and communism see, for example, Mevius, "Reappraising Communism." On consumer needs and legitimacy see Paulina Bren and Mary Neuburger, introduction to *Communism Unwrapped*, 8.

5. For an earlier work see Toomre, "Food and National Identity." For a more recent work see Bracewell, "Eating Up Yugoslavia." See also Scott, "Edible Ethnicity."

6. For some early works on alcohol and anthropology under socialism see articles in "Ethnography, Alcohol, and South-Central European Societies," special issue, *East European Quarterly* 4 (Winter 1984). See also Tulbure, "Socialist Clearing House." On Russia see, for example, Schrad, *Vodka Politics*; Trommelen and Stephenson, *Davai!*; Christian, *Living Water*; Phillips, *Bolsheviks and the Bottle*.

7. On how the concept of the "good life" appeared under Stalin in the 1930s, and the place of champagne in that notion, see Gronow, *Caviar with Champagne*. For a look at the complex notion of the good life in the Bulgarian context see Scarboro, *Late Socialist Good Life*.

8. Vazov, *Pod igoto*, 91. The translation is a combination of several available and my own.

9. For more on this see Neuburger, "Savoring the Past?"

10. See Borislavov, *Bulgarian Wine Book*, 74–79; Nikola Aretov, "Kafeneta, krŭchmi," 60.

11. Gavrilova, *Koleloto na zhivota*, 303.

12. Gavrilova, 305.

13. Gavrilova, 303–4.

14. Crampton, *Bulgaria*, 283.

15. Popov, *Stopanska Bŭlgariia*, 145.

16. See Vakarelski, *Etnografiia na Bŭlgariia*, 222–26.

17. League of Nations, *European Conference on Rural Life*, 20–21.

18. Lukacs, *Inventing Wine*.

19. See, for example, Brockett, *Bogomils of Bulgaria*.

20. Shechter, *Smoking, Culture and Economy*, 17.

21. For more on this see chapter 1 in Neuburger, *Balkan Smoke*.

22. In the modern period there is a distinct tradition of "Westerners" going "east," including to the Balkan region, and providing deeply sensory musings on food and drink that shocked and titillated audiences back home. There is a rich secondary literature analyzing these Westerners traveling east, though food and drink are rarely looked at in depth. See, for example, Todorova, *Imagining the Balkans*. A notable exception is the coverage of food and drink in travel writing on nineteenth-century southeastern Europe in Jezernik, *Wild Europe*, 47–55, 147–70.

23. Anderson, *History of the Missions*, 2:190. See also *Missionary News from Bulgaria* (Samokov), no. 21 (December 31, 1888): 6.

24. Webster, *Puritans in the Balkans*, 88.

25. Anderson, *History of the Missions*, 2:190.

26. St. Clair and Brophy, *Residence in Bulgaria*, 2.

27. *Missionary News from Bulgaria* (Samokov), no. 21 (December 31, 1888): 6.

28. Anderson, *History of the Missions*, 2:174, 181.

29. *Missionary News from Bulgaria* (Samokov), no. 21 (December 31, 1888): 6.

30. *Zornitsa* was a weekly paper published in Istanbul that was said to be circulated in more than three hundred towns in the Bulgarian provinces. By 1882 the circulation was about four thousand. Though the population was largely illiterate, the paper, like others, was regularly read aloud in Balkan cafés, and was actually mentioned in Vazov's *Under the Yoke*, 51–52, attesting to its notoriety. See Webster, *Puritans in the Balkans*, 22, 82.

31. *Report of the Seventh Convention*, 54.

32. Clarke, *Temperance Work in Bulgaria*, 3.

33. Davis, *Bulgaria Mission*, 23.

34. Stoĭkov, *Dvizhenieto za trezvenost*, 6.

35. MacDermott, *Apostle of Freedom*, 258.

36. As cited in MacDermott, 269.

37. Botev, *Sŭbrani sŭchinenie*, 2:168.

38. Slaveĭkov, *Gotvarska kniga na Diado Slaveĭkov*, 117–48. For the original publication see Slaveĭkov, *Gotvarska kniga* (1870).

39. Takhov, *Golemite bŭlgarski senzatsii*.

40. See, for example, alarm about this phenomenon in *Trezvenost* (Sofia), May 22, 1922, 7.

41. Robinson, *Oxford Companion to Wine*, 113–14.

42. Genov, *S fakela na trezvenostta*, 54; and Petkov, *Borbata za trezvenost*, 49.

43. Stoĭkov, *Dvizhenieto za trezvenost*, 7.

44. TsIDA, f. 1027k, op. 1, a.e. 46, l. 5. See also Haskell, *American Influence in Bulgaria*, 7.

45. Petkov, *Borbata za trezvenost*, 16–21.

46. See Kamenov, "Social Medicine or Racial Hygiene?"

47. The commune was named Yasna Polyana, a Bulgarian translation of the name of Tolstoy's estate, Yasnaya Polyana. For more on this see Konstantinov, *L. N. Tolstoĭ*.

48. This included Evdokia, the sister of Tsar Boris III, and two of the tsar's closest advisers, Liubomir Lulchev and Ivan Bagrianov (and presumably the tsar himself). See Kostentseva, *Moiat roden grad*, 183.

49. On alcohol and temperance in the Soviet Union see Phillips, *Bolsheviks and the Bottle*. See also Transchel, *Under the Influence*.

50. See Gronow, *Caviar with Champagne*, and Schrad, *Vodka Politics*, 225.

51. Dimitrov, *Diary*, 58, 66, 132–33.

52. Petkov, *Borbata za trezvenost*, 134.

53. On the *krŭchmari* see Znepolski, *Bŭlgarskiiat komunizŭm*, 121.

54. The law also prohibited *krŭchmi* within two hundred meters of mines, schools, factories, military bases, *chitalishte* (reading rooms or cultural centers), and military clubs.

55. Stoĭkov, *Dvizhenieto za trezvenost*, 56.

56. *Lozarstvo i vinarstvo* 12, no. 8 (1963): 24–25.

57. *Lozarstvo i vinarstvo* 11, no. 1 (1962): 2.

58. TsIDA, f. 172, op. 3, a.e. 153, l. 1; TsIDA, f. 172, op. 1, a.e. 61, ll. 15–19.

59. See, for example, *Lozarstvo i vinarstvo* 13, no. 1 (1964): 4; Zakhariev and Devedzhiev, *Teritorialno razpredelenie*, 5; and Katerov et al., *Viticulture*, 7.

60. See, for example, David Crowley and Susan Reid, introduction in *Pleasures in Socialism*, 8. See also Fehér, Heller, and Márkus, *Dictatorship over Needs*, 98.

61. See Crowley and Reid, "Style and Socialism," 10.

62. Katerov et al., *Viticulture*, 7.

63. *Lozarstvo i vinarstvo* 13, no. 1 (1964): 4.

64. *Lozarstvo i vinarstvo* 11, no. 2 (1962): 1.

65. *Lozarstvo i vinarstvo* 13, no. 1 (1964): 5.

66. On the United States see Gilby, *Wines of Bulgaria*, 26.

67. Robinson, *Oxford Companion to Wine*, 113–14.

68. Tsakov, *Vinoto*, 38–39.

69. Gilby, *Wines of Bulgaria*, 26.

70. Gilby, 27.

71. It is difficult to check such a claim, but according to Robinson, *Oxford Companion to Wine*, 114, Bulgaria reached only fourth in bottled wine exports. For the Bulgarian claim see Donchev and Katerov, *75 godini Institut po lozarstvo i vinarstvo*, 14; Katerov et al., *Viticulture*, 13.

72. Spahni, *International Wine Trade*, 299–300.

73. TsDA, f. 130k, op. 23, a.e. 300, ll. 6, 10, 19.

74. Robinson, *Oxford Companion to Wine*, 114.

75. *Lozarstvo i vinarstvo* 12, no. 7 (1963): 21.

76. *Lozarstvo i vinarstvo* 25, no. 7 (1976): 1–10.

77. *Bulgarian Foreign Trade*, March–April 1971, 27–28.

78. Fitzpatrick, *Everyday Stalinism*, 15–16.

79. See, for example, Bankov, *Kniga za vseki den*, 715–16.

80. I am indebted to Bracewell, "Eating Up Yugoslavia," for this insight pertaining to Yugoslavia. As she also notes, the same thing could apply to the West in this period.

81. Scarboro, *Late Socialist Good Life*, 172–256; Koenker, *Club Red*, 282.

82. Neuburger, "Dining in Utopia."

83. Ghodsee, *Red Riviera*. If one looks outside the Bloc, socialist Yugoslavia had an even more developed tourist economy, which can be explained by its relative openness to the West, but also by its beautiful Adriatic coast. On Yugoslav tourism see, for example, Grandits and Taylor, *Yugoslavia's Sunny Side*.

84. See, for example, Tsentralno statistichesko upravlenie, *Mezhdunaroden i vŭtreshen turizŭm*, 70, 141.

85. According to Balkanturist, the number of foreign visitors rose to one million in 1965, and to three million by 1972. Sŭbev, *90 godini organizirano turistichesko dvizhenie*, 11.

86. Markov, *Bulgarian Temptations*. The 1980s book was also published in German and French editions, with a later Bulgarian edition, *Bŭlgarski izkusheniia*, published in 1993.

87. Markov, *Bulgarian Temptations*, 24.

88. Markov, 50.

89. *Lozarstvo i vinarstvo* 11, no. 8 (1962): 1–3.

90. Zakhariev and Devedzhiev, *Teritorialno razpredelenie*, 6.

91. See, for example, Katerov et al., *Viticulture*, 5.

92. *Bulgarian Foreign Trade*, March–April 1971, 27–28.

93. *Lozarstvo i vinarstvo* 30, no. 3 (1981): 44.

94. *Lozarstvo i vinarstvo* 33, no. 4 (1984): 41; Zaĭkov et al., *Kniga za vinoto*, 42.

95. Zaĭkov et al., *Kniga za vinoto*, 23.

96. Zaĭkov et al., 46.

97. Zaĭkov et al., 47–48.

98. Apostolov, *Opiianiavashtiiat vrag*, 68.

99. Stoĭkov, *Dvizhenieto za trezvenost*, 67–68.

100. Kolarova-Paneva, Videnova, and Ivanchev, *Partiĭni i dŭrzhavni dokumenti*, 3–7, 66–67.

101. Dŭrzhaven Arkhiv Sofiia [hereafter DAS], f. 494, op. 1, a.e. 33, l. 3; Apostolov, *Opiianiavashtiiat vrag*, 20–21; Naĭdenov, *Borbata za trezvenost*, 35.

102. DAS, f. 494, op. 1, a.e. 33, l. 6; Apostolov, *Opiianiavashtiiat vrag*, 69.

103. *Trezvenost*, July 21, 1972, 2. For another example see *Trezvenost*, August 18, 1972, 1.

104. *Trezvenost*, January 10, 1975, 2; *Trezvenost*, July 1, 1972, 4.

105. *Turist*, August 1964, 15.

106. *Trezvenost*, January 1, 1975, 4.

107. DAS, f. 494, op. 1, a.e. 33, ll. 5–6. Liutov et al., *Upravlenie na narodnoto potreble-nie*, 86. See also Vlakhova-Nikolova, *Problemi na tiutiunopusheneto*, 154.

108. See, for example, Kolarova-Paneva, Videnova, and Ivanchev, *Partiĭni i dŭrzhavni dokumenti*, 6–7.

109. See Naĭdenov, *Borbata za trezvenost*, 38.

110. See Apostolov, *Opiianiavashtiiat vrag*, 66–68.

111. Nedialkov, *Bŭlgarski kulturni deĭtsii*, 11; Apostolov, *Opiianiavashtiiat vrag*, 68.

112. Nedialkov, *Bŭlgarski kulturni deĭtsii*, 7–9.

113. Nedialkov, 9–10.

114. Genov, *S fakela na trezvenostta*.

115. For an excellent article on Zhivkova see Atanasova, "Lyudmila Zhivkova."

116. Fol, *Trakiĭskiiat Dionis*, 329.

117. See, for example, Zhivkova, *Lyudmila Zhivkova*, 118.

118. Miladinova, *Esteticheska kultura*.

119. See, for example, Atanasova, "Lyudmila Zhivkova," 280–84.

120. Robinson, *Oxford Companion to Wine*, 114.

121. FAOSTAT, Food and Agriculture Organization of the United Nations, accessed January 2020, http://www.fao.org/faostat/en/#compare.

122. FAOSTAT, Food and Agriculture Organization of the United Nations, accessed January 2020, http://www.fao.org/faostat/en/#compare.

Conclusion

1. Fernández-Armesto, *Near a Thousand Tables*, xi.

BIBLIOGRAPHY

Archives

Dŭrzhaven Arkhiv Sofiia (DAS)
Lyndon Baines Johnson Presidential Library
Plovdiv Okrŭzhen Dŭrzhaven Arkhiv (PODA)
Tsentralen Istoricheski Dŭrzhaven Arkhiv (TsIDA)
United Nations Archive, UNRRA Collection

Periodicals

Bulgarian Foreign Trade
Khrana i zhivot
Lozarstvo i vinarstvo
Missionary News from Bulgaria (Samokov)
Trezvenost (Sofia)
Turist
Turizŭm
Vegetarianski pregled

Books and Articles

Albala, Ken, ed. *Routledge International Handbook of Food Studies*. London: Routledge, 2013.

Albrecht, P. "Vegetarianstvo i novata nauka za khranene." *Vegetarianski pregled* 6, no. 3 (November 1924): 41–42.

Aleksandrov, Emil. *Kultura i lichna vlast: Az rabotikh s Liudmila Zhivkova*. Sofia: Izdatelstvo "Slŭntse," 1991.

Andelman, David. "An Oil Shock of Their Own Hits Members of Comecon." *New York Times*, August 12, 1979, E4.

Anderson, Heather Arndt. *Chillies: A Global History*. London: Reaktion, 2016.

Anderson, Rufus. *History of the Missions of the American Board of Commissioners for Foreign Missions to the Oriental Churches*. Boston: Congregational Publishing Society, 1872.

Andreev, Mikhail. *Istoriia na Bŭlgarskata burzhoazna dŭrzhava i pravo: 1878–1917*. Sofia: Nauka i izkustvo, 1993.

Andreĭchin, Stefan. *Ezikŭt na vegetarianstvo*. Sofia: "Posrednik," 1940.

Andreĭchin, Stefan. "Kŭm dŭlboko osŭznavane na nashite zadachi." *Vegetarianski pregled* 8, no. 6–7 (February–March 1927): 121–23.

Apostolov, Miladin. *Opiianiavashtiiat vrag.* Sofia: n.p., 1962.

Appadurai, Arjun. "How to Make a National Cuisine: Cookbooks in Contemporary India." *Comparative Studies in Society and History* 30, no. 1 (1988): 3–24.

Appadurai, Arjun. *The Social Life of Things: Commodities in Cultural Perspective.* Cambridge: Cambridge University Press, 1988.

Aretov, Nikola. "Kafeneta, krŭchmi, saloni, i khanove v bŭlgarskata literatura ot vtorata polovina na XIX v." In *Kafene Evropa*, edited by Raia Zaimova, 60–65. Sofia: Izdatelstvo Damian Iakov, 2007.

Armstrong, Hamilton. *Tito and Goliath.* New York: Macmillan, 1951.

Artan, Tülay. "Aspects of the Ottoman Elite's Food Consumption: Looking for 'Staples,' 'Luxuries,' and 'Delicacies' in a Changing Century." In *Consumption Studies and the History of the Ottoman Empire, 1550–1922: An Introduction*, edited by Donald Quataert, 107–200. Albany: SUNY Press, 2000.

Ashley, Bob, Joanne Hollows, Steve Jones, and Ben Taylor. *Food and Cultural Studies.* London: Routledge, 2004.

Atanasova, Ivanka. "Lyudmila Zhivkova and the Paradox of Ideology and Identity in Communist Bulgaria." *East European Politics & Societies* 1 (May 2004): 278–315.

Bakić-Hayden, Milica. "Nesting Orientalisms: The Case of Former Yugoslavia." *Slavic Review* 5 (1995): 917–31.

Bankov, Minko, ed. *Kniga za vseki den i vseki dom.* Sofia: Dŭrzhavno izdatelstvo "Tekhnika," 1973.

Barkley, Henry C. *Bulgaria before the War during Seven Years' Experience of European Turkey and Its Inhabitants.* London: J. Murray, 1877.

Belasco, Warren. *Food: The Key Concepts.* Oxford: Berg, 2008.

Belchov, Todor. "Iz borbite na tiutiunorabotnitsite v grad Plovdiv (1891–1920 g.)." *Godishnik na museite v Plovdivski okrŭg* 2 (1956): 11–37.

Bell, John. *Peasants in Power: Alexander Stamboliski and the Bulgarian Agrarian National Union, 1899–1923.* Princeton, NJ: Princeton University Press, 2019.

Belorechki, Aleksandŭr. *Vegetarianstvo i surovoiadstvo.* Sofia: Meditsina i fizkultura, 1980.

Belorechki, Aleksandŭr, and Nikolaĭ A. Dzhelepov. *Mlechna kukhnia.* Sofia: Tekhnika, 1969.

Blagov, Krum. *Zagadkata Liudmila Zhivkova.* Sofia: Izdatelstvo "Reporter," 2012.

Bobrow-Strain, Aaron. *White Bread: A Social History of the Store-Bought Loaf.* Boston: Beacon, 2012.

Borislavov, Yassen. *Bulgarian Wine Book: History, Culture, Cellars, Wines.* Sofia: Trud, 2007.

Botev, Khristo. *Sŭbrani sŭchineniia.* Sofia: Bŭlgarski pisatel, 1971.

Bowden, Jonny, and Stephen T. Sinatra. *The Great Cholesterol Myth: Why Lowering Your Cholesterol Won't Prevent Heart Disease and the Statin-Free Plan That Will.* Beverly, MA: Fair Winds, 2012.

Boym, Svetlana. *The Future of Nostalgia.* New York: Basic Books, 2001.

Bozhilova, Zheni. "Novo poznanie za khranene." *Vegetarianski pregled* 12, no. 4 (December 1930): 60–62.

Bozukova, Liuba, and Zakhari Zakhariev. *Domashno konservirane na plodove i zelen-chutsi.* Sofia: Profizdat, 1959.

Bracewell, Wendy. "Eating Up Yugoslavia: Cookbooks and Consumption in Socialist Yugoslavia." In *Communism Unwrapped: Consumption in Cold War Eastern Europe,* edited by Paulina Bren and Mary Neuburger, 169–96. New York: Oxford University Press, 2013.

Bracewell, Wendy, and Alex Drace-Francis, eds. *Under Eastern Eyes: A Comparative Introduction to East European Travel Writing on Europe.* Budapest: Central European University Press, 2008.

Brandt, Karl. *Management of Agriculture and Food in the German-Occupied and Other Areas of Fortress Europe: A Study in Military Government.* Stanford, CA: Stanford University Press, 1965.

Bren, Paulina, and Mary Neuburger, eds. *Communism Unwrapped: Consumption in Cold War Eastern Europe.* Oxford: Oxford University Press, 2012.

Brockett, L. P. *The Bogomils of Bulgaria and Bosnia; or, The Early Protestants of the East: An Attempt to Restore Some Lost Leaves of Protestant History.* Philadelphia: American Baptist Publication Society, 1879.

Brulotte, Ronda L., and Michael A. Di Giovine, eds. *Edible Identities: Food as Cultural Heritage.* Burlington, VT: Ashgate, 2014.

Bunin, Ivan. *The Liberation of Tolstoy: A Tale of Two Writers.* Edited and translated by Thomas Gaiton Marullo and Vladimir T. Khmelkov. Evanston, IL: Northwestern University Press, 2001.

Bunin, Ivan, Thomas Alekseevich, Marullo Gaiton, and Vladimir Khmelkov. *The Liberation of Tolstoy: A Tale of Two Writers.* Evanston, IL: Northwestern University Press, 2001.

Burlikov, Dimo. "Blagosŭstoianie i khranene." *Khrana i zhivot* 2, no. 7 (March 1941): 121.

Caldwell, Melissa, ed. *Food and Everyday Life in the Post-Socialist World.* Bloomington: Indiana University Press, 2009.

Carolan, Michael S. *The Real Cost of Cheap Food.* London: Earthscan, 2011.

Castle, Shushana, and Amy-Lee Goodman. *The Meaty Truth: Why Our Food Is Destroying Our Health and Environment—and Who Is Responsible.* New York: Skyhorse, 2014.

Castro, Josué de. *The Geography of Hunger.* Boston: Little, Brown, 1952.

Centre for Co-operation with Non-Members and Organisation for Economic Co-operation and Development. *Review of Agricultural Policies.* Paris: Organisation for Economic Co-operation and Development, 2000.

Chatterjee, Choi. *Celebrating Women: Gender, Festival Culture, and Bolshevik Ideology.* Pittsburgh: University of Pittsburgh Press, 2002.

Chevat, Richie, and Michael Pollan. *The Omnivore's Dilemma: The Secrets behind What You Eat.* New York: Penguin, 2007.

Chirot, Daniel. "Social Change in Communist Romania." *Social Forces* 57 (1978): 457–99.

Cholcheva, Penka, Sonia Chortanova, and Nadezhda Ilieva. "Umeem da gotvim." In *Kniga za vseki den i vseki dom,* edited by Minko Bankov, 408–554. Sofia: Dŭrzhavno izdatelstvo "Tekhnika," 1973.

Chortanova, Sonia, and Nikolaĭ Dzhelepov. *Nashata i svetovnata kukhnia i ratsional-noto khranene*. Sofia: Meditsina i fizkultura, 1977.

Christian, David. *Living Water: Vodka and Russian Society on the Eve of Emancipation*. Oxford: Clarendon, 1990.

Clarke, James F. *Temperance Work in Bulgaria: Its Successes*. Samokov: Evangelical School Press, 1909.

Cochrane, N., A. Schmitz, and S. Bojnec. "Agriculture: Diversification and Productivity." In *Privatization of Agriculture in New Market Economies: Lessons from Bulgaria*, edited by Andrew Schmitz, Kirby Moulton, Allan Buckwell, and Sofia Davidova, 23–53. New York: Springer, 1994.

Cochrane, Nancy. *Hard Currency Constraints and East European Grain Imports*. Washington, DC: US Department of Agriculture, Economic Research Service, Agriculture and Trade Analysis Division, 1988.

Collar, C. "Bread: Types of Bread." In *Encyclopedia of Food and Health*, edited by Benjamin Caballero, Paul M. Finglas, and Fidel Toldrá, 1:500–507. Oxford: Academic Press, 2016.

Collingham, Lizzie. *The Taste of War: World War II and the Battle for Food*. New York: Penguin, 2012.

Connelly, John. "Nazis and Slavs: From Racial Theory to Racist Practice." *Central European History* 32, no. 1 (1999): 1–33.

Crampton, Richard J. *Bulgaria*. Oxford: Oxford University Press, 2007.

Crampton, Richard J. *Eastern Europe in the Twentieth Century and After*. London: Routledge, 1997.

Crampton, Richard J. "The Turks in Bulgaria, 1878–1944." In *The Turks of Bulgaria: The History and Fate of a Minority*, edited by K. H. Karpat, 43–78. Istanbul: Isis, 1990.

Crawford, Beverly. *Economic Vulnerability in International Relations: The Case of East-West Trade, Investment, and Finance*. New York: Columbia University Press, 1993.

Crowley, David, and Susan E. Reid, eds. *Pleasures in Socialism: Leisure and Luxury in the Eastern Bloc*. Evanston, IL: Northwestern University Press, 2010.

Crowley, David, and Susan E. Reid. "Style and Socialism: Modernity and Material Culture." In *Style and Socialism: Modernity and Material Culture in Postwar Eastern Europe*, edited by David Crowley and Susan E. Reid, 1–24. Oxford: Berg, 2000.

Crowley, David, and Susan E. Reid. *Style and Socialism: Modernity and Material Culture in Postwar Eastern Europe*. Oxford: Berg, 2000.

Currie, Hugh M. *The Flying Scotsman: Collected Thoughts on Thirteen Years of Travel, 1964–1977*. Campbell River, BC: Burslem, 2004.

Cwiertka, Katarzyna. *Modern Japanese Cuisine: Food, Power and National Identity*. London: Reaktion, 2007.

Dande-Vansel, A. "Belite brashna, lisheni ot zaroditsa na zhitoto." *Vegetarianski pregled* 12, no. 6 (February 1931): 91–93.

Daskalov, Rumen. *Bŭlgarskoto obshtestvo, 1878–1939*. Vol. 1, *Dŭrzhava, politika, ikonomika*. Sofia: IK "Gutenberg," 2005.

Daskalov, Rumen. *Mezhdu iztoka i zapada: Bŭlgarski kulturni dilemi*. Sofia: Lik, 1998.

Davis, Dora. *The Bulgaria Mission of the Methodist Episcopal Church*. New York: Missionary Society of the Methodist Episcopal Church, 1906.

Davydov, A. I. *Nelegal'noe snabzhenie rossiiskogo naseleniia i vlast' 1917–1921 gg.: Meshochniki*. St. Petersburg: "Nauka," 2002.

Debré, Patrice. *Louis Pasteur*. Translated by Elborg Forster. Baltimore: Johns Hopkins University Press, 1998.

Demireva, Mara. *Izmeneniia v khraneneto na bŭlgarskiia narod*. Sofia: Izdatelstvo na Bŭlgarskata akademiia na naukite, 1968.

Desai, Padma. *Estimates of Soviet Grain Imports in 1980–85: Alternative Approaches*. Washington, DC: International Food Policy Research Institute, 1981.

Detchev, Stefan. "From Istanbul to Sarajevo via Belgrade—A Bulgarian Cookbook of 1874." In *Earthly Delights: Economies and Cultures of Food in Ottoman and Danubian Europe, c. 1500–1900*, edited by Angela Jianu and Violeta Barbu, 376–401. Leiden: Brill, 2018.

Detchev, Stefan. "'Liutite chushki' i bŭlgarskata natsionalna kukhnia: Mezhdu istoricheskata realnost i sotsialnoto vŭobrazhenie." *Istorichesko bŭdeshte* 1–2 (2013): 59–107.

Detchev, Stefan. "Mezhdu vishata osmanska kukhnia i Evropa: Slaveĭkovata kniga ot 1870 g. i pŭtiat kŭm modernoto gotvarstvo." *Littera et Lingua* 11, no. 3 (2014). https://naum.slav.uni-sofia.bg/lilijournal/2014/11/3/dechevs.

Detchev, Stefan. "Shopska Salad: From a European Innovation to the National Culinary Symbol." In *From Kebab to Ćevapčići: Foodways in (Post-)Ottoman Europe*, edited by Arkadiusz Blaszczyk and Stefan Rohdewald, 273–88. Wiesbaden: Harrassowitz Verlag, 2018.

Detchev, Stefan. "'Shopska salata': Kak se razhda edin natsionalen kulinaren simvol." In *V tŭrsene na bŭlgarskoto: Mrezhi na natsionalna intimnost (XIX–XXI vek)*, edited by Stefan Detchev, 411–63. Sofia: Institut za izsledvane na izkustvata, 2010.

Deutsch, Robert. *The Food Revolution in the Soviet Union and Eastern Europe*. Boulder, CO: Westview, 1986.

Dilchev, Konstantin. *Zagadŭchnata smŭrt na Liudmila Zhivkova: Agni ĭoga—taĭnoto uchenie*. Sofia: Niu media grup, 2006.

Dimitrov, Georgi. *The Diary of Georgi Dimitrov, 1933–1949*. New Haven, CT: Yale University Press, 2008.

Dimitrov, Georgi, Mosko Raĭchev, and Stefan Stefanov. *Mekhanizatsiia i avtomatizatsiia pri proizvodstvoto na khliab i khlebni izdeliia*. Sofia: Tekhnika, 1966.

Dimkov, Petŭr. *Bŭlgarska narodna meditsina: Prirodolechenie i prirodosŭobrazen zhivot*. Sofia: Izdatelstvo na Bŭlgarskata akademiia na naukite, 1977.

Dimou, Augusta. *Entangled Paths towards Modernity: Contextualizing Nationalism and Socialism in the Balkans*. Budapest: Central European University Press, 2009.

Dimov, Nikola. *Mlekarstvo i mlekokontrola*. Sofia: Zemizdat, 1949.

Donchev, A., and Kaliu Katerov. *75 godini Institut po lozarstvo i vinarstvo: Nositel na ordenite Georgi Dimitrov i N.R. Bŭlgariia; Pleven 1902–1977*. Pleven: Institut po lozarstvo i vinarstvo, 1977.

Dosev, Khristo. *Blizo do Iasna Poliana, 1907–1909 g*. Sofia: Lingua optima consilium, 2010.

Dosev, Khristo. *Etika na khranata, ili Pioneri na vegetarianstvo*. Burgas, Bulgaria: Izdanie na vŭzrazhdane, 1911.

Douglas, Loudon M. *The Bacillus of Long Life*. New York: G. P. Putnam's Sons, 1911.

Douno, Beinsa. *Food and Water: Messages from Heaven*. Lexington, KY: Astrala, 2013.

Dragoĭcheva, Tsola. *Defeat to Victory: Notes of a Bulgarian Revolutionary*. Sofia: Sofia Press, 1983.

Dragoĭcheva, Tsola, and Stefan Zhelev. *Povelia na dŭlga: Spomeni i razmisli*. Sofia: Partizdat, 1972.

Dragostinova, Theodora. *The Cold War from the Margins: A Small Socialist State on the Global Cultural Scene*. Ithaca, NY: Cornell University Press, 2021.

Drews, G. "Zashto ne deistvuvashe vegetarianstvoto do sega." *Vegetarianski pregled* 12, no. 9 (May 1931): 133–35.

Dzhelepov, Nikolai, and Aleksandŭr Belorechki. "Dietichno khranene u doma pri razlichni zaboliavaniia." In *Kniga za vseki den i vseki dom*, edited by Minko Bankov, 302–29. Sofia: Tekhnika, 1973.

Edgerton, William. "The Social Influence of Lev Tolstoj in Bulgaria." In *American Contributions to the Tenth International Congress of Slavists, Sofia 1988*, edited by Jane Harris, 123–38. Columbus, OH: Slavica, 1988.

Elenkov, Ivan. "Versii za bŭlgarskata identichnost v modernata epokha." In Ivan Elenkov and Rumen Daskalov, *Zashto sme takiva? V tŭrsene na bŭlgarskata kulturna identichnost*, 5–26. Sofia: Prosveta, 1994.

Elenkov, Ivan, and Rumen Daskalov. *Zashto sme takiva? V tŭrsene na bŭlgarskata kulturna identichnost*. Sofia: Prosveta, 1994.

Elias, Norbert. *The Civilizing Process*. New York: Urizen Books, 1978.

Emanuilov, Ignat. *Veterinarno-sanitarna ekspertiza na khranitelnite produkti ot zhivotinski proizkhod*. Sofia: Lito-pechat "Trud," 1946.

Embree, George Daniel. *The Soviet Union between the 19th and 20th Party Congresses, 1952–1956*. The Hague: Nijhoff, 1959.

Estabrook, Barry. *Tomatoland: How Modern Industrial Agriculture Destroyed Our Most Alluring Fruit*. Kansas City: Andrews McMeel, 2011.

"Ethnography, Alcohol, and South-Central European Societies." Special issue, *East European Quarterly* 18, no. 4 (Winter 1984).

Fehér, Ferenc, Ágnes Heller, and György Márkus. *Dictatorship over Needs: An Analysis of Soviet Societies*. New York: St. Martin's, 1983.

Fermor, Patrick Leigh. *The Broken Road*. New York: New York Review of Books, 2013.

Fernández-Armesto, Felipe. *Civilizations: Culture, Ambition, and the Transformation of Nature*. New York: Free Press, 2001.

Fernández-Armesto, Felipe. *Near a Thousand Tables: A History of Food*. New York: Free Press, 2002.

Fitzpatrick, Sheila. *Everyday Stalinism: Ordinary Life in Extraordinary Times; Soviet Russia in the 1930s*. New York: Oxford University Press, 1999.

Fol, Aleksandŭr. *Trakiĭskiiat Dionis*. Sofia: Universitetsko izdatelstvo "Sv. Kliment Okhridski," 1991.

Fondatsiia "Dr Stamen Grigorov." *V nachaloto be rodovata pamet*. Sofia: Universitetsko izdatelstvo "Sv. Kliment Okhridski," 2005.

Fox, Frank. *Bulgaria*. London: A. and C. Black, 1915.

Frankel, Max. "Red Countries Seek U.S. Wheat: Czechoslovakia, Bulgaria and Hungary Offering to Buy 60 Million in Grain." *New York Times*, October 4, 1963, 1.

Ganov, Zakhari. "Prekhranata na natsiata e vazhen stimiul na neĭnat vŭzkhod." *Khrana i zhivot* 2, no. 7 (March 1941): 122–23.

Gavrilova, Raĭna. *Koleloto na zhivota: Vsekidnevieto na bŭlgarskiia vŭzrozhdenski grad.* Sofia: Universitetsko izdatelstvo "Sv. Kliment Okhridski," 1999.

Gavrilova, Raĭna. *Semeĭnata stsena: Antropologicheska istoriia na semeĭnoto khranene v Bŭlgariia v modernata epokha.* Sofia: Universitetsko izdatelstvo "Sv. Kliment Okhridski," 2016.

Gavrilova, Rayna. "Golden Fruits from the Orchards." In *Culinary Cultures of Europe: Identity, Diversity and Dialogue,* edited by Darra Goldstein, Kathrin Merkle, Fabio Parasecoli, and Stephen Mennell, 95–102. Strasbourg: Council of Europe Publications, 2005.

Genov, Paun. *S fakela na trezvenostta: Momenti ot borbata protiv piianstvoto i tiutiunopushteneto prez 1300-godishnata istoriia na Bŭlgariia.* Sofia: Natsionalen komitet za trezvenost, meditsina i fizkultura, 1980.

Gentilcore, David. *Pomodoro! A History of the Tomato in Italy.* New York: Columbia University Press, 2010.

Georgieff, Anthony. "The Rise and Fall of the Bulgarian Kebapche." *Vagabond,* November 1, 2008. https://vagabond.bg/index.php/rise-and-fall-bulgarian-kebapche-2315.

Georgiev, Petŭr. *Bial ili cher khliab?* Sofia: Biblioteka "Vŭzrodena Bŭlgariia," 1940.

Georgiev, V. P. *Dŭlgoletieto na choveka.* Sofia: Izdatelstvo na Natsionalniia sŭvet na Otechestveniia front, 1960.

Gerhard, Gesine. *Nazi Hunger Politics: A History of Food in the Third Reich.* Lanham, MD: Rowman & Littlefield, 2015.

Geshov, Ivan. *Nashite gradinari druzhestva.* Sofia: n.p., 1889.

Ghodsee, Kristen. *Lost in Transition: Ethnographies of Everyday Life after Communism.* Durham, NC: Duke University Press, 2011.

Ghodsee, Kristen. "Red Nostalgia? Communism, Women's Emancipation, and Economic Transformation in Bulgaria." *L'Homme: Zeitschrift für feministische Geschichtswissenschaft* 15 (2004): 33–46.

Ghodsee, Kristen. *The Red Riviera: Gender, Tourism, and Postsocialism on the Black Sea.* Durham, NC: Duke University Press, 2005.

Gilby, Caroline. *The Wines of Bulgaria, Romania and Moldova.* Oxford: Infinite Ideas, 2018.

Glants, Musya, and Joyce Toomre, eds. *Food in Russian History and Culture.* Bloomington: Indiana University Press, 1997.

Gocheva, Rositsa. *Razvitie na materialnoto blagosŭstoianie na bŭlgarskiia narod.* Sofia: Izdatelstvo na Bŭlgarskata komunisticheska partiia, 1965.

Goldstein, Darra, and Kathrin Merkle, eds. *Culinary Cultures of Europe: Identity, Diversity and Dialogue.* Cologne: Lübbe, 2005.

Golovinski, Evgeni. *Osŭshtestvenata mechta na Asen Zlatarov.* Sofia: Dŭrzhavno izdatelstvo "Nauka i izkustvo," 1981.

Grandits, Hannes, and Karin Taylor, eds. *Yugoslavia's Sunny Side: A History of Tourism in Socialism, 1950–1980.* Budapest: Central European Press, 2010.

Grayzel, Susan. *Women and the First World War*. New York: Routledge, 2013.

Gronow, Jukka. *Caviar with Champagne: Common Luxury and the Ideals of the Good Life in Stalin's Russia*. New York: Berg, 2003.

Groueff, Stéphane. *Crown of Thorns*. Lanham, MD: Madison Books, 1987.

Gruev, Mikhail. *Preorani slogove: Kolektivizatsiia i sotsialna promiana v Bŭlgarskiia severo-zapad, 40-te–50-te godini na XX vek*. Sofia: Ciela, 2009.

Haskell, Arnold. *Heroes and Roses: A View of Bulgaria*. London: Darton, Longman & Todd, 1966.

Haskell, Edward. *American Influence in Bulgaria*. New York: American Board of Commissioners for Foreign Missionaries, 1913.

Hearings before the Committee on Foreign Relations, United States Senate, Eighty-Ninth Congress, First Session, on the Shipping Restrictions on Grain Sales to Eastern Europe, September 17 and 27, 1965. Washington, DC: GPO, 1965.

Herald of Health. "The Sour Milk Fetish." *Health* 60 (1910): 262–63.

Hyman, Clarissa. *Tomato: A Global History*. London: Reaktion, 2019.

Iacobbo, Karen, and Michael Iacobbo. *Vegetarian America: A History*. Westport, CT: Praeger, 2006.

İnalcık, Halil, and Donald Quataert. *An Economic and Social History of the Ottoman Empire, 1300–1914*. Cambridge: Cambridge University Press, 1994.

Iordachi, Constantin, and Arnd Bauerkämper. *The Collectivization of Agriculture in Communist Eastern Europe: Comparison and Entanglements*. Budapest: Central European University Press, 2014.

Iordanov, Todor. *Materialnoto-tekhnicheska basa na razvitoto sotsialistichesko obshtestvo*. Sofia: Partizdat, 1973.

Ĭosifov, Kalin. *Totalitarizmŭt v bŭlgarskoto selo: Khronika na nasilieto*. Sofia: Universitetsko izdatelstvo "Sv. Kliment Okhridski," 1999.

Irwin, Julia. "Taming Total War: Great War–Era American Humanitarianism and Its Legacies." In *Beyond 1917: The United States and the Global Legacies of the Great War*, edited by Thomas W. Zeiler, David K. Ekbladh, and Benjamin C. Montoya, 122–39. New York: Oxford University Press, 2017.

Işın, Priscilla Mary. *Bountiful Empire: A History of Ottoman Cuisine*. London: Reaktion Books, 2018.

Ivanov, Ivan. *Kiselo mliako: Bŭlgarskoto ime na dŭlgoletieto*. Obshtina Trŭn: Fondatsiia "Dr Stamen Grigorov," 2006.

Ivanov, Vicho. "Tagore in Bulgaria." In *Rabindranath Tagore: A Centenary Volume, 1861–1961*, 323–31. New Delhi: Sahitya Akademi, 1961.

Jedlicki, Jerzy. *A Suburb of Europe: Nineteenth Century Polish Approaches to Western Civilization*. Budapest: Central European University Press, 1999.

Jezernik, Bozidar. *Wild Europe: The Balkans in the Gaze of Western Travellers*. London: Saqi, 2003.

Johnson, Stowers. *Gay Bulgaria*. London: R. Hale, 1964.

Judt, Tony. *Postwar: A History of Europe since 1945*. New York: Penguin, 2005.

Jung, Yuson. "From Canned Food to Canny Consumers: Cultural Competence in the Age of Mechanical Production." In *Food and Everyday Life in the Post-Socialist World*, edited by Melissa L. Caldwell, 29–56. Bloomington: Indiana University Press, 2009.

Kamenov, Nikolay. "A Question of Social Medicine or Racial Hygiene? The Bulgarian Temperance Discourse and Eugenics in the Interwar Period, 1920–1940." In *Global Anti-vice Activism, 1890–1950: Fighting Drinks, Drugs, and "Immorality,"* edited by Jessica R. Pliley, Robert Kramm, and Harald Fischer-Tiné, 124–51. Cambridge: Cambridge University Press, 2016.

Kaneva-Johnson, Maria. *The Melting Pot: Balkan Food and Cookery.* London: Prospect Books, 1994.

Kantardzhiev, Asen. *Mlekarski narechnik.* Sofia: Pridvorna pechatnitsa, 1930.

Karakasheva, A. *Gotvarska kniga.* Sofia: n.p., 1930.

Karaosmanoğlu, Defne. "Cooking the Past: The Revival of Ottoman Cuisine." PhD diss., McGill University, 2006.

Karavelov, Liuben. *Zapiski za Bŭlgariia i za bŭlgarite.* Sofia: Dŭrzhavna pechatnitsa, 1930.

Karavelov, Liuben. "Zapiski za Bŭlgariia i za bŭlgarite." In *Vŭzrozhdenski pŭtepisi,* edited by Svetla Giurova, 167–246. Sofia: Bŭlgarski pisatel, 1969.

Karpat, K. H., ed. *Turks of Bulgaria: The History and Fate of a Minority.* Istanbul: Isis, 1990.

Kaser, Karl, ed. *Household and Family in the Balkans: Two Decades of Historical Family Research at University of Graz.* Berlin: Lit Verlag, 2012.

Kassabova, Kapka. *Street without a Name: Childhood and Other Misadventures in Bulgaria.* New York: Skyhorse, 2009.

Kastelov, Boian. *Bŭlgariia: Ot voǐna kŭm vŭstanie.* Sofia: Voenno izdatelstvo, 1988.

Katerov, Kaliu, P. Mamarov, I. Chalkov, and Konstantina Stoyanova. *Viticulture and Wine Industry Development in the People's Republic of Bulgaria.* Sofia: Academy of Agricultural Sciences, Bulgaria, Center for Scientific, Technical and Economic Information in Agriculture and Forestry, 1971.

Katrandzhiev, K. *Bŭlgarskoto kiselo mliako: Mikrobiologichni i dietichni kachestva.* Sofia: BAN, 1962.

Kauffman, Jonathan. *Hippy Food: How Back-to-the-Landers, Longhairs, and Revolutionaries Changed the Way We Eat.* New York: William Morrow, 2018.

Kellogg, John H. *The Battle Creek Sanitarium: History, Organization, Methods.* Battle Creek, MI: Self-published, 1913.

Kendall, Arthur. "Recent Developments in Intestinal Bacteriology." *American Journal of Medical Sciences* 156 (August 1918): 157–73.

Kennedy, John F. *The Kennedy Presidential Press Conferences.* New York: E. M. Coleman, 1978.

Kennedy, John F. "Letter to the President of the Senate and to the Speaker of the House on the Sale of Wheat to the Soviet Union," October 10, 1963. American Presidency Project, online by Gerhard Peters and John T. Woolley. https://www.presidency.ucsb.edu/documents/letter-the-president-the-senate-and-the-speaker-the-house-the-sale-wheat-the-soviet-union.

Khadzhiĭski, Ivan, and Mariia Khadzhiĭska. *Bit i dushevnost na nashiia narod.* Sofia: Lik izdaniia, 2002.

Khadzhiĭski, Tŭrpo, Nikola Pekachev, Pinkas Koen, Dimitŭr Donkov, and Margarita Tsolova. *Domashno konservirane.* Sofia: Zemizdat, 1976.

Khrushchev, Nikita, and Sergei Khrushchev. *Memoirs of Nikita Khrushchev.* Vol. 3, *Statesman, 1953–1964.* University Park: Penn State University Press, 2007.

Kiossev, Alexander. "The Self-Colonizing Metaphor." In *Atlas of Transformation* (online). http://monumenttotransformation.org/atlas-of-transformation/html/s/self-colonization/the-self-colonizing-metaphor-alexander-kiossev.html.

Kitanina, Tasiia, and S. I. Potolov. *Voina, khleb i revoliutsiia: Prodovol'stvennyi vopros v Rossii, 1914–Oktiabr' 1917 g.* Leningrad: Izdatel'stvo "Nauka," Leningradskoe otdelenie, 1985.

Kiumzhiev, Ivan. "Kŭm delo!" *Vegetarianski pregled* 5, no. 1 (September 1923): 1–3.

Kiumzhiev, Ivan. "Shtadete zhivota!" *Vegetarianski pregled* 5, no. 2 (October 1923): 17–19.

Koenker, Diane. *Club Red: Vacation Travel and the Soviet Dream.* Ithaca, NY: Cornell University Press, 2013.

Kolarova-Paneva, Dimitrina, Fani Videnova, and Spas Ivanchev. *Partiĭni i dŭrzhavni dokumenti po trezvenostta.* Sofia: Meditsina i fizkultura, 1984.

Kondratenko, Maria. *Bŭlgarsko kiselo mliako.* Sofia: Zemizdat, 1985.

Konstantinov, Aleko. *Bai Ganyo: Incredible Tales of a Modern Bulgarian.* Translated by Victor A. Friedman, Christina E. Kramer, Grace E. Fielder, and Catherine Rudin. Madison: University of Wisconsin Press, 2010.

Konstantinov, Aleko. *Do Chikago i nazad.* In vol. 1 of *Sŭbrani sŭchineniia,* edited by Tikhomir Tikhov. Sofia: Bŭlgarski Pisatel, 1980.

Konstantinov, Aleko. *To Chicago and Back.* Translated by Robert Sturm. Sofia: National Museum of Bulgarian Books and Polygraphy, 2004.

Konstantinov, Georgi. *L. N. Tolstoĭ i vliianieto mu v Bŭlgariia.* Sofia: Biblioteka "Svobodna misŭl," 1968.

Konstantinova, Nevyana. *Zov za budnost na sŭrtsata.* lulu.com, 2010.

Kostentseva, Raĭna. *Moiat roden grad Sofiia predi 75 godini i posle.* Sofia: Izdatelstvo na Otechestveniia front, 1979.

Kostentseva, Raĭna. *Moiat roden grad Sofiia v kraia na XIX–nachalo na XX vek i sled tova.* Sofia: "Riva," 2008.

Kulchytsky, Stanislav. *The Famine of 1932–1933 in Ukraine: An Anatomy of the Holodomor.* Translated by Ali Kinsella. Toronto: Canadian Institute of Ukrainian Studies Press, 2018.

Kulp, Walter L., and Leo F. Rettger. "Comparative Study of *Lactobacillus acidophilus* and *Lactobacillus bulgaricus.*" *Journal of Bacteriology* 9, no. 4 (1924): 357–95.

Lakhtikova, Anastasia, Angela Brintlinger, and Irina Glushchenko. *Seasoned Socialism: Gender and Food in Late Soviet Everyday Life.* Bloomington: Indiana University Press, 2019.

Lampe, John. *The Bulgarian Economy in the Twentieth Century.* New York: St. Martin's, 1986.

Lampe, John. "Imperial Borderlands or Capitalist Periphery? Redefining Balkan Backwardness, 1520–1914." In *The Origins of Backwardness in Eastern Europe: Economics and Politics from the Middle Ages until the Early Twentieth Century,* edited by Daniel Chirot, 177–209. Berkeley: University of California Press, 1989.

Lampe, John, and Marvin Jackson. *Balkan Economic History, 1550–1950: From Imperial Borderlands to Developing Nations.* Bloomington: Indiana University Press, 1982.

Lang, George. *The Cuisine of Hungary.* New York: Bonanza Books, 1971.

Laudan, Rachel. *Cuisine and Empire: Cooking in World History*. Berkeley: University of California Press, 2013.

League of Nations. *European Conference on Rural Life: National Monographs Drawn Up by Governments; Bulgaria*. Geneva: League of Nations, 1940.

League of Nations. *The Problem of Nutrition*. Geneva: League of Nations, 1936.

LeBlanc, Ronald. "The Ethics and Politics of Diet: Tolstoy, Pilnyak, and the Modern Slaughterhouse." *Gastronomica* 17, no. 4 (2017): 9–25.

LeBlanc, Ronald. *Slavic Sins of the Flesh: Food, Sex, and Carnal Appetite in Nineteenth-Century Russian Fiction*. Durham: University of New Hampshire Press, 2009.

Lemke, Thomas. *Biopolitics: An Advanced Introduction*. Translated by Eric Frederick Trump. New York: NYU Press, 2011.

Leslie, Henrietta. *Where East Is West: Life in Bulgaria*. Boston: Houghton Mifflin, 1933.

Lih, Lars. *Bread and Authority in Russia, 1914–1921*. Berkeley: University of California Press, 1990.

Liutov, Atanas, Boris Atanasov, Violeta Samardzhieva, and Katia Stoianova. *Upravlenie na narodnoto potreblenie*. Sofia: Izdatelstvo na Bŭlgarskata akademiia na naukite—Ikonomicheski institut, 1984.

Long, Lucy, ed. *Culinary Tourism*. Lexington: University of Kentucky Press, 2004.

Lozinski, E. "Mesoiadie i prestŭpnost." *Vegetarianski pregled* 8, no. 8–9 (April–May 1923): 154–55.

Luif, Paul. *Security in Central and Eastern Europe: Problems, Perceptions, Policies*. Vienna: Braumüller, 2001.

Lukacs, Paul. *Inventing Wine: A New History of One of the World's Most Ancient Pleasures*. New York: W. W. Norton, 2012.

MacDermott, Mercia. *The Apostle of Freedom: A Portrait of Vasil Levsky against a Background of Nineteenth Century Bulgaria*. South Brunswick, NJ: A. S. Barnes, 1969.

MacDermott, Mercia. *Bulgarian Folk Customs*. Philadelphia: Jessica Kingsley, 1998.

Marinov, Dimitŭr. "Material za Bŭlgarskiia rechnik: Dumi i frazi iz zapadna Bŭlgariia." In *Sbornik za narodni umotvoreniia, nauka i knizhina*, 13:249–71. Sofia: Ministerstvo na narodnoto prosveshtenie, 1896.

Marinov, Tchavdar. "Ancient Thrace in the Modern Imagination: Ideological Aspects of the Construction of Thracian Studies in Southeast Europe (Romania, Greece, Bulgaria)." In *Entangled Histories of the Balkans*, vol. 3, *Shared Pasts, Disputed Legacies*, edited by Roumen Daskalov and Alexander Vezenkov, 10–117. Leiden: Brill, 2013.

Markov, Emil. *Bulgarian Temptations: 33 Illustrated Culinary Journeys with Recipes*. Sofia: Balkanturist, 1981.

Markov, Emil. *Bŭlgarski izkusheniia: 33 iliustrovani kulinarni pŭteshestviia, 330 gotvarski retsepti*. Sofia: Petrum Ko, 1993.

McCauley, Martin. *Khrushchev and the Development of Soviet Agriculture: The Virgin Land Programme, 1953–1964*. Teaneck, NJ: Holmes & Meier, 1976.

McCauley, Martin. *The Khrushchev Era, 1953–1964*. Harlow, UK: Longman / Pearson Education, 1995.

McGrath, Maria. *Food for Dissent: Natural Foods and the Consumer Counterculture since the 1960s*. Amherst: University of Massachusetts Press, 2019.

McLeod, Michael J., Sheldon I. Guttman, and W. Hardy Eshbaugh. "Peppers." In *Isozymes in Plant Genetics and Breeding, Part B*, edited by S. D. Tanksley and T. J. Orton, 189–201. Amsterdam: Elsevier Science, 1983.

Megyeri, Zsofia. "Bulgarian Yoghurt Market Dynamics in 2012." *Progressive*, October 21, 2013. http://progressive.bg/en/magazine/on-focus/bulgarian-yoghurt-market-dynamics-in-2012/1447/.

Mendelson, Anne. *Milk: The Surprising Story of Milk through the Ages, with 120 Adventurous Recipes That Explore the Riches of Our First Food.* New York: Knopf, 2008.

Mennell, Stephen. "On the Civilizing of Appetite." *Theory, Culture and Society* 4, no. 2–3 (1987): 373–403.

Metchnikoff, Élie. *The Prolongation of Life: Optimistic Studies.* 1907. Reprint, New York: G. P. Putnam's Sons, 1912.

"Metchnikoff's Theory of Longevity." *Scientific American*, May 16, 1908, 347.

Mevius, Martin. "Reappraising Communism and Nationalism." In *The Communist Quest for National Legitimacy in Europe, 1918–1989*, edited by Martin Mevius, 1–24. New York: Routledge, 2011.

Mezhdunarodna konferentsiia po vitaminite: Dokladi i sŭobshteniia. Sofia: BAN, 1962.

Miladinova, Mila. *Esteticheska kultura i trezvenost.* Sofia: Meditsina i fizkultura, 1979.

Minchev, Tsoniu. *Kratko praktichesko rŭkovodstvo za mlekonadzora v mandrite.* Sofia: Tsentralen kooperativen sŭiuz, 1948.

Mishkova, Diana. *Beyond Balkanism: The Scholarly Politics of Region Making.* New York: Routledge, 2020.

Mlekuž, Jernej. *Burek: A Culinary Metaphor.* Budapest: Central European University Press, 2015.

Mocheva, Khristina. *Selskoto zemedělsko domakinstvo v Bŭlgariia prez 1935/36 godina: Biudzhet, obstanovka i razkhod na trud.* Sofia: Dŭrzhavna pechatnitsa, 1938.

Montanari, Massimo. *Italian Identity in the Kitchen, or Food and the Nation.* New York: Columbia University Press, 2013.

"Most Obese Countries." Reuters, October 5, 2010. https://www.reuters.com/news/picture/most-obese-countries-idUSRTXT3DK.

Murdzhev, Dimitŭr. *Taka gi vidiakh: Dŭlgi godini v okhranata na Zhivkovi, smŭrtta na L. Zhivkova, lovets N 1, tankovete na Mladenov, Vanga i oshte.* Sofia: Poligraficheski kombinat, 1992.

Nabhan, Gary Paul. *Where Our Food Comes From: Retracing Nikolay Vavilov's Quest to End Famine.* Washington, DC: Island Press / Shearwater Books, 2009.

Naĭdenov, Ivan, and Sonia Chortanova. *Nasha kukhnia.* Sofia: Nauka i izkustvo, 1955.

Naĭdenov, Ivan, and Sonia Chortanova. *Rŭkovodstvo po obshtestveno khranene.* Sofia: Nauka i izkustvo, 1953.

Naĭdenov, Ivan, and Ivan Zakhariev. *Prouchvane organizatsiiata na mlekoproizvoditelnite fermi v raiona na Sofiia.* Sofia: Bŭlgarskata akademiia na naukite, 1960.

Naĭdenov, Vasil. *Borbata za trezvenost—delo na tseliia narod: Materiali ot okrŭzhnata nauchnoprakticheska konferentsiia po problemite na trezvenostta, provedena prez maĭ 1979 g.* Plovdiv: Okrŭzhen komitet za trezvenost—Plovdiv, 1980.

Nauchno-konsultativen sŭvet kŭm AMB. *Bŭlgarskoto ime na dŭlgoletieto: 100 godini ot otkrivaneto na Lactobacillus bulgaricus.* Sofia: SP Betaprint, 2005.

Nedialkov, Khristo. *Bŭlgarski kulturni deĭtsii za trezvenost.* Sofia: Meditsina i fizkultura, 1977.

Neuburger, Mary. *Balkan Smoke: Tobacco and the Making of Modern Bulgaria, 1863–1989*. Ithaca, NY: Cornell University Press, 2012.

Neuburger, Mary. "Dining in Utopia: A Taste of the Bulgarian Black Sea Coast under Socialism." *Gastronomica* 17, no. 4 (Fall 2017): 48–60.

Neuburger, Mary. *The Orient Within: Muslim Minorities and the Negotiation of Nationhood in Modern Bulgaria*. Ithaca, NY: Cornell University Press, 2004.

Neuburger, Mary. "Savoring the Past? Food and Drink in Nineteenth-Century Narratives on Ottoman and Post-Ottoman Bulgaria." In *From Kebab to Ćevapčići: Foodways in (Post-) Ottoman Europe*, edited by Arkadiusz Blaszczyk and Stefan Rohdewald, 257–72. Wiesbaden: Harrassowitz Verlag, 2018.

Neumann, Iver. *Russia and the Idea of Europe: A Study in Identity and International Relations*. New York: Routledge, 1996.

New York Times. "Economic Growth Brisk in Red Bloc." March 20, 1967.

New York Times. "Metchnikoff Confirmed in His Theory of Long Life." January 21, 1912.

Nimmo, Richie. *Milk, Modernity, and the Making of the Human: Purifying the Social*. London: Routledge, 2010.

Noncheva, Theodora Ivanovna. *The Winding Road to the Market: Transition and the Situation of Children in Bulgaria*. Florence: UNICEF International Child Development Centre, 1995.

Nova gotvarska kniga. Sofia: Vestnik na zhenata, n.d.

Ogle, Maureen. *In Meat We Trust: An Unexpected History of Carnivore America*. New York: Houghton Mifflin Harcourt, 2013.

Oren, Nissan. *Revolution Administered: Agrarianism and Communism in Bulgaria*. Baltimore: Johns Hopkins University Press, 1973.

Organisation for Economic Co-operation and Development. *OECD Review of Agricultural Policies: Bulgaria 2000*. Paris: Organisation for Economic Co-operation and Development, 2000.

Palairet, Michael. *The Balkan Economies: Evolution without Development*. Cambridge: Cambridge University Press, 2003.

Partŭchev, Vladimir Khristov. *Nashite stoletnitsi: Statistiko-biologichesko izsledvane*. Sofia: Glavna direktsi na statistika, 1933.

Patel, Raj. *Stuffed and Starved: The Hidden Battle for the World Food System*. New York: Melville House, 2008.

Patenaude, Bertrand M. *The Big Show in Bololand: The American Relief Expedition to Soviet Russia in the Famine of 1921*. Stanford, CA: Stanford University Press, 2002.

Patterson, Patrick. *Bought and Sold: Living and Losing the Good Life in Socialist Yugoslavia*. Ithaca, NY: Cornell University Press, 2011.

Pavlov, Ivan. *Prisŭstviia na khraneneto po bŭlgarskite zemi prez XV–XIX vek*. Sofia: Akademichno izdatelstvo "Marin Drinov," 2001.

Pells, Richard H. *Not Like Us: How Europeans Have Loved, Hated, and Transformed American Culture since World War II*. New York: Basic Books, 1997.

Peteri, Gyuri. "Nomenklatura with Smoking Guns: Hunting in Communist Hungary." In *Pleasures in Socialism: Leisure and Luxury in the Eastern Bloc*, edited by David Crowley and Susan E. Reid, 311–43. Evanston, IL: Northwestern University Press, 2010.

"Peti sŭbor na vegetarianski sŭiuz." *Vegetarianski pregled* 5, no. 1 (September 1923): 3–10.

Petkov, Pavel. *Borbata za trezvenost vŭv Vrachanski okrŭg, 1920–1980.* Sofia: Izdatelstvo na Otechestveniia front, 1982.

Petrov, Liubomir, Nikolai Dzhelepov, Evgeni Iordanov, and Snezhina Uzunova. *Bŭlgarska natsionalna kukhnia.* Sofia: Zemizdat, 1978.

Petrov, Tseno. *Agrarnite reformi v Bŭlgariia, 1880–1944.* Sofia: Izdatelstvo na Bŭlgarskata akademiia na naukite, 1975.

Petrov, Victor. "The Rose and the Lotus: Bulgarian Electronic Entanglements in India, 1967–89." *Journal of Contemporary History* 54, no. 3 (2019): 666–87.

Phillips, Laura L. *Bolsheviks and the Bottle: Drink and Worker Culture in St. Petersburg, 1900–1929.* DeKalb: Northern Illinois University Press, 2000.

Ploss, Sidney. *Conflict and Decision Making in Soviet Russia.* Princeton, NJ: Princeton University Press, 1965.

Pod znamenem Oktiabria: Sbornik dokumentov i materialov. Moscow: Izdatel'stvo politicheskoi literatury Bolgarskoi kommunisticheskoi partii, 1981.

Podolsky, Scott. "Cultural Divergence: Elie Metchnikoff's *Bacillus bulgaricus* Therapy and His Underlying Concept of Health." *Bulletin of the History of Medicine* 72, no. 1 (1998): 1–27.

Popdimitrov, K. *Bŭlgarsko kiselo mliako: Proizkhod, proizvodtsvo, khranitelnost i nadzor.* Sofia: Spas Iv. Bozhinov, 1938.

Popkin, Barry. "Nutritional Patterns and Transitions." *Population and Development Review* 19, no. 1 (1993): 138–57.

Popov, Kiril. *Stopanska Bŭlgariia prez 1911 god: Statisticheski izsledvaniia.* Sbornik na Bŭlgarskata akademiia na naukite, book 8. Sofia: Dŭrzhavna pechatnitsa, 1916.

"Pork and Bacon." *Bulgaria Today* 11 (1962): 45.

Pozharliev, Raĭcho. *Filosofiia na khraneneto: Kulturno-istoricheski konteksti.* Sofia: Universitetsko izdatelstvo "Sv. Kliment Okhridski," 2013.

Pringle, Peter. *The Murder of Nikolai Vavilov: The Story of Stalin's Persecution of One of the Great Scientists of the Twentieth Century.* New York: Simon & Schuster, 2008.

Proctor, Robert. *Racial Hygiene: Medicine under the Nazis.* Cambridge, MA: Harvard University Press, 1988.

Quammen, David. "The Bear Slayer." *Atlantic*, July/August 2003. https://www.the atlantic.com/magazine/archive/2003/07/the-bear-slayer/302768/.

Radio Free Europe. *Radio Free Europe Research.* [Washington, DC]: Radio Free Europe, 1974.

Raĭkin, Spas. *Rebel with a Just Cause: A Political Journey against the Winds of the 20th Century.* Sofia: Pensoft, 2001.

Rakovski, Georgi. *Izbrani sŭchineniia.* Sofia: Dŭrzhavno izdatelstvo, 1946.

Ram, Hari Har, and Rakesh Yadava. *Genetic Resources and Seed Enterprises: Management and Policies.* New Delhi: New India, 2007.

Report of the Seventh Convention of the World's Woman's Christian Temperance Union, Tremont Temple, Boston, Massachusetts, October 17th–23rd, 1906. Evanston, IL: The Union, 1906.

Riley, Barry. *The Political History of American Food Aid: An Uneasy Benevolence.* New York: Oxford University Press, 2017.

Robinson, Jancis, ed. *The Oxford Companion to Wine*. 4th ed. Oxford: Oxford University Press, 2015.

Salisbury, Harrison E. "Bulgaria Seeks Trade with U.S." *New York Times*, November 12, 1960, 25.

Saraliev, Petŭr. *Gotvarska kniga za mŭzhe*. Sofia: Zemizdat, 1984.

Scarboro, Cristofer. *The Late Socialist Good Life in Bulgaria: Meaning and Living in a Permanent Present Tense*. Lanham, MD: Lexington Books, 2012.

Scarboro, Cristofer, Diane Mincyte, and Zsuzsa Gille, eds. *The Socialist Good Life: Desire, Development, and Standards of Living in Eastern Europe*. Bloomington: Indiana University Press, 2020.

Scholliers, Peter, ed. *Food, Drink and Identity: Cooking, Eating and Drinking in Europe since the Middle Ages*. London: Bloomsbury Academic, 2001.

Schrad, Mark. *Vodka Politics: Alcohol, Autocracy, and the Secret History of the Russian State*. Oxford: Oxford University Press, 2014.

Schwarcz, Joseph. *An Apple a Day: The Myths, Misconceptions, and Truths about the Foods We Eat*. New York: Other Press, 2009.

Scott, Erik. "Edible Ethnicity: How Georgian Cuisine Conquered the Soviet Table." *Kritika* 13 (2012): 831–58.

Semerdzhiev, Petŭr. *Narodniiat sŭd v Bŭlgariia, 1944–1945 g.: Komu i zashto e bil neobkhodim*. Sofia: Makedoniia Press, 1998.

Semov, Mincho, and Ivanka Iankova. *Bŭlgarskite gradove prez Vŭzrazhdaneto: Istorichesko, sotsiologichesko i politologichesko izsledvane*. Sofia: Universitetsko izdatelstvo "Sv. Kliment Okhridski," 2004.

Seppain, Hélène. *Contrasting US and German Attitudes to Soviet Trade, 1917–91: Politics by Economic Means*. Hampshire, UK: Macmillan, 1992.

Sharkey, Heather. *A History of Muslims, Christians, and Jews in the Middle East*. Cambridge: Cambridge University Press, 2017.

Sharp, Ingrid, and Matthew Stibbe. *The Aftermaths of War: Women's Movements and Female Activists, 1918–1923*. Leiden: Brill, 2011.

Shechter, Relli. *Smoking, Culture and Economy in the Middle East: The Egyptian Tobacco Market, 1850–2000*. New York: I. B. Tauris, 2006.

Sheimanov, Naiden. "Preobrazhenie na Bŭlgariia." In *Zashto sme takiva? V tŭrsene na bŭlgarskata kulturna identichnost*, edited by Ivan Elenkov and Rumen Daskalov, 266–69. Sofia: Prosveta, 1994.

Shepard, C. H. "What the Doctors Say: Old Age May Be Kept at Bay by the Use of Bulgarian Yoghourt." *Phrenological Journal and Science of Health* 1119 (April 1908): 117–20.

Shkodrova, Albena. "From Duty to Pleasure in the Cookbooks of Communist Bulgaria: Attitudes to Food in the Culinary Literature for Domestic Cooking Released by the State-Run Publishers between 1949 and 1989." *Food, Culture & Society* 21, no. 4 (August 2018): 468–87.

Shkodrova, Albena. *Rebellious Cooks and Recipe Writing in Communist Bulgaria*. London: Bloomsbury Academic, 2021.

Shkodrova, Albena. *Sots gurme: Kurioznata istoriia na kukhniata v NRB*. Plovdiv: Zhanet 45, 2014.

Shopov, Georgi. *Na gosti v Iasna-Poliana*. Sofia: Biblioteka "Svobodna misŭl," 1929.

Shprintzen, Adam. *The Vegetarian Crusade: The Rise of an American Reform Movement, 1817–1921*. Chapel Hill: University of North Carolina Press, 2013.

Shub, Boris, and Zorach Warhaftig, *Starvation over Europe (Made in Germany): A Documented Record, 1943*. New York: Institute of Jewish Affairs of the American Jewish Congress and World Jewish Congress, 1943.

"60% of Tomatoes on Bulgarian Market Are Imported." Novinite.com, December 3, 2018. https://www.novinite.com/articles/193646/60+of+Tomatoes+on+Bulgarian+Market+are+Imported.

Slaveĭkov, Petko. *Gotvarska kniga, ili, Nastavleniia za vsiakakvi gozbi, spored kakto gi praviat v Tsarigrad, i razni domashni spravi: Sŭbrani ot razni knigi*. Tsarigrad: Pechatnitsa "Makedoniia," 1870.

Slaveĭkov, Petko. *Gotvarska kniga na Diado Slaveĭkov*. Sofia: Millenium, 2015.

Slosson, Edwin. "Twelve Major Prophets of Today." *Independent*, December 7, 1911, 1235–50.

Smith, Andrew. *The Tomato in America: Early History, Culture, and Cookery*. Columbia: University of South Carolina Press, 1994.

Smith, Barbara Clark. "Food Rioters and the American Revolution." *William and Mary Quarterly* 51, no. 1 (1994): 3–38.

Smollett, Eleanor Wenkart. "The Economy of Jars: Kindred Relationships in Bulgaria—an Exploration." *Ethnologia Europaea* 19, no. 2 (1989): 125–40.

Sotirov, Natsko. *Sŭvremenna kukhnia: 3000 retsepta*. Sofia: Dŭrzhavno izdatelstvo "Tekhnika," 1959.

Soyfer, Valery N. *Lysenko and the Tragedy of Soviet Science*. New Brunswick, NJ: Rutgers University Press, 1994.

Spahni, Pierre. *The International Wine Trade*. Cambridge: Woodhead, 2000.

Spencer, Colin. *The Heretic's Feast: A History of Vegetarianism*. London: Fourth Estate, 1993.

"The Spiritual Master Peter Deunov." Beinsa Douno—Plovdiv. http://beinsadouno-plovdiv.org/all/english/05t.html.

St. Clair, Stanislas, and Charles A. Brophy. *A Residence in Bulgaria; or, Notes on the Resources and Administration of Turkey: The Condition and Character, Manners, Customs, and Language of the Christian and Musselman Populations, with Reference to the Eastern Question*. London: J. Murray, 1869.

St. Gavriĭski. "Otraviane s meso v Nashensko." *Vegetarianski pregled* 8, no. 10 (June 1927): 212–15.

Stanchev, N. "Khrana i khranitelnost." *Vegetarianski pregled* 8, no. 2 (September 1926): 27–35.

Stanchev, N. "Nauka na khranata." *Vegetarianski pregled* 8, no. 10 (March 1932): 180–81.

Stanford, Craig. *The Hunting Apes: Meat Eating and the Origins of Human Behavior*. Princeton, NJ: Princeton University Press, 1999.

Steinberg, Deborah Lynn. *Genes and the Bioimaginary: Science, Spectacle, Culture*. Abingdon, Oxon: Routledge, 2016.

Stevenson, David. *With Our Backs to the Wall: Victory and Defeat in 1918*. Cambridge, MA: Belknap Press of Harvard University Press, 2011.

Stillman, Edmund O. "The Collectivization of Bulgarian Agriculture." In *The Collectivization of Agriculture in Eastern Europe*, edited by Irwin T. Sanders, 67–102. Lexington: University of Kentucky Press, 1958.

Stoĭchkov, Ĭordan. *Lovnite retsepti na Bai Dancho: 100 kulinari idei za nai-vkusen divech.* Sofia: Millenium, 2012.

Stoĭkov, Khristo. *Dvizhenieto na trezvenost v Razgradski okrŭg, 1920–1982.* Razgrad, Bulgaria: Okrŭzhen komitet za trezvenost, 1983.

Stoilova, Elitsa. "From a Homemade to an Industrial Product: Manufacturing Bulgarian Yogurt." *Agricultural History* 87, no. 1 (2013): 73–92.

Stoilova, Elitsa. "Producing Bulgarian Yoghurt: Manufacturing and Exporting Authenticity." PhD diss., Eindhoven University of Technology, 2014.

Strauss, Johann. "Twenty Years in the Ottoman Capital: The Memoirs of Dr. Hristo Tanev Stambolski of Kazanlik (1843–1932) from an Ottoman Point of View." In *Istanbul—Kushta—Constantinople: Narratives of Identity in the Ottoman Capital, 1830–1930,* edited by Christoph Herzog and Richard Wittmann, 246–302. New York: Routledge, 2019.

Stuart, Tristram. *The Bloodless Revolution: A Cultural History of Vegetarianism from 1600 to Modern Times.* New York: W. W. Norton, 2007.

Sŭbev, Vicho. *90 godini organizirano turistichesko dvizhenie v Bŭlgariia.* Sofia: Meditsina i fizkultura, 1986.

Takhov, Rosen. *Golemite bŭlgarski senzatsii.* Sofia: Izdatelstvo "Iztok-Zapad," 2005.

Tashev, Tasho. *Khranene, fizichesko razvitie i zdravno sŭstoianie na Bŭlgarskiia narod.* Sofia: BAN, 1972.

Tchakarov, Kostadin. *The Second Floor: An Exposé of Backstage Politics in Bulgaria.* London: Macdonald, 1991.

Tepev, N. *Turskoto naselenie v NR Bŭlgariia: Materiali za politicheskite zaniatiia na trudovatsite.* Sofia: Dŭrzhavno voenno izdatelstvo, 1954.

Thompson, E. P. "The Moral Economy of the English Crowd in the Eighteenth Century." *Past & Present,* no. 50 (February 1971): 76–136.

Times (London). "Science and Sour Milk." April 29, 1910, 11.

Todorov, Goran. "Deĭnostta na Vremennoto rusko upravlenie v Bŭlgariia po urezhdane na agrarniia i bezhanskiia vŭpros prez 1877–1879 g." *Istoricheski pregled* 6 (1955): 54–59.

Todorov, Pŭrvan. *Golemiiat khliab: Osnovateli na kooperativnoto zemedelsko stopanstvo "Vŭzkhod" v selo Totleben, Plevensko, razkazvat za nachaloto, postaveno prez 1939 godina.* Sofia: Zemizdat, 1988.

Todorova, Maria. *Imagining the Balkans.* New York: Oxford University Press, 1997.

Todorova, Maria, and Zsuzsa Gille, eds. *Post-Communist Nostalgia.* London: Berghahn Books, 2012.

Tolstoy, Leo. *The First Step: An Essay on the Morals of Diet, to Which Are Added Two Stories.* Manchester: Albert Broadbent, 1900.

Tomov, Ivan. *Neshta, koito triabva da znaem pri khraneneto si.* Sofia: Biblioteka "fond Khimik pri Sŭiuza na Bŭlgarskite khimitsi," 1948.

Toomre, Joyce. "Food and National Identity in Soviet Armenia." In *Food in Russian History and Culture,* edited by Musya Glants and Joyce Toomre, 195–214. Bloomington: Indiana University Press, 1997.

Toussaint-Samat, Maguelonne. *The History of Food.* Translated by Anthea Bell. New York: Blackwell, 1992.

Trager, James. *The Great Grain Robbery.* New York: Ballantine Books, 1975.

Transchel, Kate. *Under the Influence: Working-Class Drinking, Temperance, and Cultural Revolution in Russia, 1895–1932.* Pittsburgh: University of Pittsburgh Press, 2006.

Trentmann, Frank, and Flemming Just, eds. *Food and Conflict in Europe in the Age of the Two World Wars.* Hampshire, UK: Palgrave Macmillan, 2006.

Trommelen, Edwin, and David Stephenson. *Davai! The Russians and Their Vodka.* Montpelier, VT: Russian Life Books, 2012.

Trotsky, Leon. *The Balkan Wars, 1912–13: The War Correspondence of Leon Trotsky.* Edited by George Weissman and Duncan Williams. Translated by Brian Pearce. New York: Pathfinder, 1980.

Tsachev, Khristo, Sofiia Ioncheva, and Magdalena Mladenova. *Ot vйrkha na XX vek: Minalo, nastoiashte i bйdeshte na mesoprerabotvatelnata industriia v Bйlgariia.* Sofia: Kooperativno-izdatelska kйshta "KHBP," 1999.

Tsachev, Khristo, and Georgi Stoĭchev. *Mesopromishlenostta v Bйrgaski krai.* Sofia: Profizdat, 1988.

Tsakov, Dimitйr. *Vinoto: Traditsiia, kultura, sйvremennost.* Sofia: Stefka Georgieva, 2008.

Tsentralno statistichesko upravlenie pri ministerskiia sйvet. *Mezhdunaroden i vйtreshen turizйm, 1960–1967.* Sofia: Tsentralno statistichesko upravlenie, 1968.

Tulbure, Narcis. "The Socialist Clearing House: Alcohol, Reputation, and Gender in Romania's Second Economy." In *Communism Unwrapped: Consumption in Cold War Eastern Europe,* edited by Paulina Bren and Mary Neuburger, 255–76. New York: Oxford University Press, 2013.

Turncock, David. *East European Economy in Context: Communism and Transition.* New York: Routledge, 1997.

United Nations Interim Commission on Food and Agriculture. *The Work of FAO: A General Report to the First Session of the Conference of the Food and Agriculture Organization of the United Nations.* Washington, DC: United Nations Interim Commission on Food and Agriculture, 1945.

"Uses of Sour Milk." *Youth's Companion* 84 (October 20, 1910): 552.

Vakarelski, Khristo. *Etnografiia na Bйlgariia.* Sofia: Izdatelstvo na nauka i izkustvo, 1974.

Valenze, Deborah. *Milk: A Local and Global History.* New Haven, CT: Yale University Press, 2011.

Vangelov, Georgi. *Tiutiunorabotnitsi: Spomeni iz belezhki iz borbite na tiutiunorabotnitsite v Bйlgariia.* Sofia: Profizdat, 1955.

Vatralski, Stoian. "Koi i kakvi sa Belite bratia." In *Izgrevйt na Bialoto bratstvo, pee i sviri, uchi i zhivee,* edited by Vergiliĭ Krйstev, 24:191–235. Sofia: Biblioteka "Zhiten klas," 2008.

Vazov, Ivan. *The Great Rila Wilderness.* Sofia: Sofia Press, 1969.

Vazov, Ivan. *Pod igoto.* Sofia: Bйlgarski pisatel, 1956.

Vazov, Ivan. *Under the Yoke.* Translated by Marguerite Alexieva and Theodora Atanassova. New York: Twayne, 1971.

Vazov, Ivan. *Under the Yoke.* Translated by Edmund Gosse. London: Heinemann, 1894.

"Vegetarianska kukhnia." *Vegetarianski pregled* 12, no. 7 (March 1931): 109–10.

Veit, Helen Z. *Modern Food, Moral Food: Self-Control, Science, and the Rise of Modern American Eating in the Early Twentieth Century.* Chapel Hill: University of North Carolina Press, 2013.

Velichkov, Petйr. *Kakvo khapnakha i piĭnakha velikite bйlgari.* Sofia: Izdatelstvo ERA, 2013.

Vlakhova-Nikolova, Veselina. *Problemi na tiutiunopusheneto i alkokholnata upotreba sred mladezhta*. Plovdiv: Nauchnoizsledovatelska laboratoriia za mladezhta, 1983.

V. M. "Protein." *Vegetarianski pregled* 8, no. 10 (March 1932): 182–83.

Volkov, Nikolai. "The Aftermaths of Defeat: The Fallen, the Catastrophe, and the Public Response of Women to the End of the First World War in Bulgaria." In *The Aftermaths of War: Women's Movements and Female Activists, 1918–1923*, edited by Ingrid Sharp and Matthew Stibbe, 29–47. Leiden: Brill, 2011.

Von Bremzen, Anya. *Mastering the Art of Soviet Cooking: A Memoir of Food and Longing*. New York: Crown, 2013.

Wädekin, Karl Eugen, and Everett M. Jacobs. *Agrarian Policies in Communist Europe: A Critical Introduction*. Totowa, NJ: Allanheld, Osmun, 1982.

Walters, Kerry, and Lisa Portmess. *Ethical Vegetarianism: From Pythagoras to Peter Singer*. Albany: SUNY Press, 1999.

Walton, Stuart. *The Devil's Dinner: A Gastronomic and Cultural History of Chili Peppers*. New York: St. Martin's, 2018.

Webster, William. *Puritans in the Balkans: The American Board Mission in Bulgaria, 1878–1918; A Study in Purpose and Procedure*. Sofia: Studia Historico-Philologica Serdicensia, 1938.

Wiley, Andrea. *Re-imagining Milk: Cultural and Biological Perspectives*. New York: Routledge, 2011.

Wilson, Thomas M. "Food, Drink and Identity in Europe: Consumption and the Construction of Local, National and Cosmopolitan Culture." In *Food, Drink and Identity in Europe*, edited by Thomas M. Wilson, 11–29. Amsterdam: Rodopi, 2006.

Wood, Elizabeth. *The Baba and the Comrade: Gender and Politics in Revolutionary Russia*. Bloomington: Indiana University Press, 1997.

Wrangham, Richard. *Catching Fire: How Cooking Made Us Human*. New York: Basic Books, 2009.

Yotova, Maria. "'It Is the Bacillus That Makes Our Milk': Ethnocentric Perceptions of Yogurt in Postsocialist Bulgaria." In *Why We Eat, How We Eat: Contemporary Encounters between Foods and Bodies*, edited by Emma-Jayne Abbots and Anna Lavis, 169–85. New York: Routledge, 2013.

Yotova, Maria. "Reflecting Authenticity: 'Grandmother's Yogurt' between Bulgaria and Japan." In *Food between the Country and the City: Ethnographies of a Changing Global Foodscape*, edited by Nuno Domingos, José Manuel Sobral, and Harry G. West, 175–90. New York: Bloomsbury Academic, 2014.

Zaharieva, Virginia. *Nine Rabbits*. Translated by Angela Rodel. New York: Black Balloon, 2014.

Zaĭkov, Iliia, Ivan Dionisiev, Georgi Petrov, and Kosta Forev. *Kniga za vinoto*. Sofia: n.p., 1982.

Zakhariev, Ivan, and Marin Devedzhiev. *Teritorialno razpredelenie i ikonomicheska efektivnost na vinarskata promishlenost v NRB*. Sofia: BAN, 1969.

Zaraska, Marta. *Meathooked: The History and Science of Our 2.5-Million-Year Love Affair with Meat*. New York: Basic Books, 2015.

"Zelenchutsi i salati." *Vegetarianski pregled* 8, no. 8–9 (April–May 1927): 193–94.

Zhekova, Mira. *Shto e ratsionalna khranene*. Sofia: Meditsina i fizkutura, 1982.

Zhivkova, Liudmila. *Lyudmila Zhivkova: Her Many Worlds, New Culture and Beauty, Concepts and Action*. Oxford: Pergamon, 1986.

Zlatarov, Asen. "Gladuvashta Bŭlgariia." *Chervena tribuna* 4, no. 4 (April 1932): 78–80.

Zlatarov, Asen. *Izbrani proizvedeniia*. Sofia: Nauka i izkustvo, 1975.

Zlatarov, Asen. *Izbrani sŭchineniia v tri toma*. Sofia: Nauka i izkustvo, 1966.

Zlatarov, Asen. *Osnovi na naukata za khraneneto: Lektsii, dŭrzhani prez uchebnata 1920–1921 godina v Sofiĭskiia universitet*. Sofia: Paskalev, 1921.

Zlatarov, Asen. *Osnovi na naukata za khraneneto: Lektsii, dŭrzhani prez uchebnata 1920–1921 godina v Sofiĭskiia universitet*. 2nd ed. Sofia: Izdatelstvo "Bŭlgarska knizhnitsa," 2016.

Znamenski, Andrei. *Red Shambhala: Magic, Prophecy, and Geopolitics in the Heart of Asia*. Wheaton, IL: Theosophical Publishing House, 2011.

Znepolski, Ivaĭlo. *Bŭlgarskiiat komunizŭm: Sotsiokulturni cherti i vlastova traektoriia*. Sofia: Ciela, 2008.

Znepolski, Ivaylo, Mihail Gruev, Momtchil Metodiev, Martin Ivanov, Daniel Vatchkov, Ivan Elenkov, and Plamen Doynov. *Bulgaria under Communism*. New York: Routledge, 2019.

❧ Index

Lightning Source UK Ltd.
Milton Keynes UK
UKHW012134080322
399764UK00004B/153